Sea Surface Temperature: From Observation to Applications

Sea Surface Temperature: From Observation to Applications

Editor

Francisco Pastor

MDPI • Basel • Beijing • Wuhan • Barcelona • Belgrade • Manchester • Tokyo • Cluj • Tianjin

Editor
Francisco Pastor
Fundación CEAM
Spain

Editorial Office
MDPI
St. Alban-Anlage 66
4052 Basel, Switzerland

This is a reprint of articles from the Special Issue published online in the open access journal *Journal of Marine Science and Engineering* (ISSN 2077-1312) (available at: https://www.mdpi.com/journal/jmse/special_issues/sea_temperature).

For citation purposes, cite each article independently as indicated on the article page online and as indicated below:

LastName, A.A.; LastName, B.B.; LastName, C.C. Article Title. *Journal Name* **Year**, *Volume Number*, Page Range.

ISBN 978-3-0365-2600-3 (Hbk)
ISBN 978-3-0365-2601-0 (PDF)

Cover image courtesy of Francisco Pastor

Contents

Preface to "Sea Surface Temperature: From Observation to Applications"

In a global and accelerated climate change environment, the sea surface temperature (SST) was defined by the World Meteorological Organization as one of the essential climate variables that contribute to the characterization of the Earth's climate. Recent studies have confirmed that a huge amount of energy is being stored in the oceans; therefore, the SST has emerged as a proxy of this energy reservoir, especially to derive future trends in climate change and the impacts on the frequency of weather extremes and their growing impact on human societies. This energy storage has a considerable impact on the atmosphere–ocean system through heat exchange.

In recent years, more attention has been paid to SST monitoring and analysis and, as a result, notable advances have been recorded in these fields. This book looks at these recent advances in SST studies regarding issues such as the SST–hurricane relationship, regional warming trends and future climate scenarios.

As Guest Editor, I would like to thank the authors of the manuscripts in this book, and I would also like to thank the effort and work of the MDPI team who have made editing this book such an easy and enriching experience.

Francisco Pastor
Editor

Journal of
*Marine Science
and Engineering*

Editorial

Sea Surface Temperature: From Observation to Applications

Francisco Pastor

Meteorology and Pollutant Dynamics Area, Fundación CEAM, Paterna, 46980 Valencia, Spain; paco@ceam.es

Citation: Pastor, F. Sea Surface
Temperature: From Observation to
Applications. *J. Mar. Sci. Eng.* **2021**, *9*,
1284. https://doi.org/10.3390/
jmse9111284

Received: 2 November 2021
Accepted: 17 November 2021
Published: 18 November 2021

Publisher's Note: MDPI stays neutral
with regard to jurisdictional claims in
published maps and institutional affiliations.

Sea surface temperature (SST) has been defined by World Meteorological Organization (WMO) as one of the essential climate variables (ECVs) contributing to the characterization of Earth's climate. As one of the ECVs, SST study and analysis have been receiving growing interest in the last two decades, especially as new databases from satellites have become increasingly available with higher spatial and temporal resolution. In a global and accelerated climate change environment, SST can be understood as a proxy of the ocean's role as an energy storage facility. This role is especially important to derive future trends in climate change and their impacts on climate, weather extremes, marine ecosystems, and on human societies.

The main mechanism by which the ocean interacts with the atmosphere is through heat and moisture exchanges. Those processes exert a major influence on the development of extreme weather events, such as hurricanes or torrential rains, which are projected to increase in frequency and intensity in the Sixth Assessment Report of the Intergovernmental Panel on Climate Change (IPCC) [1]. Hence, good knowledge of SST patterns and trends is crucial to investigate interactions/feedback with the atmosphere, climate drivers such as El Niño, and marine biodiversity, but especially to understand predicted future climate scenarios. Some of these issues have been addressed in this Special Issue, such as marine heat waves (MHW), SST, and hurricanes interaction or SST trends and projections. This assessment has been mainly done by analyzing data measured from satellites, but proper validation with in-situ data is necessary to test and validate satellite sensors. This has also been addressed in this Special Issue in one of the most biologically diverse and under menace ecosystems on Earth, the coral reefs.

Satellites measure skin and subskin temperature at the very surface of the sea in a thin layer at 10 µm and 1 mm depth. Other science branches, such as marine biology, may need to know temperature at different depths, so the feasibility of using SST in those studies needs to be investigated. In [2], Gómez et al. investigate the correlation between different satellite SST databases and in-situ data at the coral reef depth. They found a good correlation between satellite and in-situ measurements, expected as coral reefs lie under shallow waters in coastal subtidal areas, but with some seasonal bias that can be used to correct satellite data for its use in coral reef surveillance. In a somewhat similar work [3], Colin et al. look for a correlation between a temporally long and spatially extensive temperature monitoring network, at different sea depths and going deeper than the previous reference, and satellite SST and sea surface height (SSH) data. The authors were able to create a regression model with SST and SSH capable of predicting depth-varying thermal stress from satellite measurements. These two studies highlight the need for and complementary of both types of SST measurement, in situ and satellite. This highlights the importance of promoting both technological development of measurement methodologies and calibration and validation studies.

Regarding extreme weather events, [4] investigates the relationship between a highly active hurricane season in the Pacific and the SST and upper ocean heat content (UOHC). They found good relationship between both SST and UOHC anomalies and intensity of three major hurricanes, and with the spatial extent of hurricane tracks. These results reinforce the interest of the deepening of knowledge in such ocean–atmosphere energy exchanges that trigger extreme weather phenomena that are expected to increase in the coming decades.

In addition, it is important to analyze SST actual trends and projected scenarios in the framework of climate change. For the Adriatic Sea, [5] detect a positive trend for both SST and air temperature, but with a lag between them. The authors also point to the impact that the long-term SST warming trend in the Adriatic has already had in marine fauna and the implications of climate change in the Adriatic islands population and development. At a greater scale, climate shaping phenomena like El Niño events can also be studied through SST analysis, among other variables, as shown in [6]. Once actual and past phenomena and trends are increasingly well known, the focus should be on analysing future climate scenarios, among which the evolution of the SST is of particular interest in this Special Issue. Indeed, [7] studied decadal and seasonal SST variation in the East China Shelf seas using Couple Model Intercomparison Project (CMIP) models under RCP4.5, demonstrating an expected 1.5 °C increase by 2100. This SST increase, especially the seasonal one, can play an important role in future climate and marine ecology changes in the area.

In recent years, a poorly studied phenomenon has been gaining insight in ocean science literature—marine heat waves (MHWs). MHWs are defined as a prolonged discrete anomalously warm water event that can be described by its duration, intensity, rate of evolution, and spatial coverage. MHWs impact on marine biodiversity is and will be a major topic of study in the coming years. Refs. [8,9] study the recent evolution of MHWs in the Mediterranean and Red Seas, where their frequency and intensity have clearly increased in the last four decades. Finally, [10] analyzes the main characteristics of MHWs in the Red Sea.

As is evident from the variety of topics covered in the articles of this Special Issue, sea surface temperature is involved in a wide range of fields and topics of great relevance to marine and climate science. SST can be used as a tool for analysis in many fields, from extreme weather events to marine biology, so it is of great interest both to deepen our understanding of it, and to improve the way we approach it and how we measure it. Although a high level of excellence has been achieved in the measurement of SST, especially in its measurement from satellites, there is undoubtedly much progress to be made in this field.

Advancing knowledge about SST will certainly be of great help in analysing, understanding, and trying to cope with the already observed and expected future climate change on Earth, and we encourage and request all researchers in this field to maintain and redouble their efforts, in the interest of science and humankind.

Funding: This research received no external funding.

Institutional Review Board Statement: Not applicable.

Informed Consent Statement: Not applicable.

Data Availability Statement: Not applicable.

Conflicts of Interest: The authors declare no conflict of interest.

References

1. IPCC. *2021: Climate Change 2021: The Physical Science Basis. Contribution of Working Group I to the Sixth Assessment Report of the Intergovernmental Panel on Climate Change*; Masson-Delmotte, V.P., Zhai, A., Pirani, S.L., Connors, C., Péan, S., Berger, N., Caud, Y., Chen, L., Goldfarb, M.I., Gomis, M., et al., Eds.; Cambridge University Press: Cambridge, UK, 2021; in press.
2. Gomez, A.M.; McDonald, K.C.; Shein, K.; DeVries, S.; Armstrong, R.A.; Hernandez, W.J.; Carlo, M. Comparison of Satellite-Based Sea Surface Temperature to In Situ Observations Surrounding Coral Reefs in La Parguera, Puerto Rico. *J. Mar. Sci. Eng.* **2020**, *8*, 453. [CrossRef]
3. Colin, P.L.; Johnston, T.M.S. Measuring Temperature in Coral Reef Environments: Experience, Lessons, and Results from Palau. *J. Mar. Sci. Eng.* **2020**, *8*, 680. [CrossRef]
4. Ford, V.L.; Walker, N.D.; Pun, I.-F. Anomalous Oceanic Conditions in the Central and Eastern North Pacific Ocean during the 2014 Hurricane Season and Relationships to Three Major Hurricanes. *J. Mar. Sci. Eng.* **2020**, *8*, 288. [CrossRef]

5. Bonacci, O.; Bonacci, D.; Patekar, M.; Pola, M. Increasing Trends in Air and Sea Surface Temperature in the Central Adriatic Sea (Croatia). *J. Mar. Sci. Eng.* **2021**, *9*, 358. [CrossRef]
6. Wijaya, Y.J.; Wisha, U.J.; Hisaki, Y. The North Equatorial Countercurrent East of the Dateline, Its Variations and Its Relationship to the El Niño Event. *J. Mar. Sci. Eng.* **2021**, *9*, 1041. [CrossRef]
7. Lu, H.; Xie, C.; Zhang, C.; Zhai, J. CMIP5-Based Projection of Decadal and Seasonal Sea Surface Temperature Variations in East China Shelf Seas. *J. Mar. Sci. Eng.* **2021**, *9*, 367. [CrossRef]
8. Ibrahim, O.; Mohamed, B.; Nagy, H. Spatial Variability and Trends of Marine Heat Waves in the Eastern Mediterranean Sea over 39 Years. *J. Mar. Sci. Eng.* **2021**, *9*, 643. [CrossRef]
9. Mohamed, B.; Nagy, H.; Ibrahim, O. Spatiotemporal Variability and Trends of Marine Heat Waves in the Red Sea over 38 Years. *J. Mar. Sci. Eng.* **2021**, *9*, 842. [CrossRef]
10. Bawadekji, A.; Tonbol, K.; Ghazouani, N.; Becheikh, N.; Shaltout, M. General and Local Characteristics of Current Marine Heatwave in the Red Sea. *J. Mar. Sci. Eng.* **2021**, *9*, 1048. [CrossRef]

Journal of
*Marine Science
and Engineering*

Article

Measuring Temperature in Coral Reef Environments: Experience, Lessons, and Results from Palau

Patrick L. Colin [1],* and T. M. Shaun Johnston [2]

1 Coral Reef Research Foundation, P.O. Box 1765, Koror 96940, Palau
2 Scripps Institution of Oceanography, University of California, San Diego, La Jolla, CA 92093, USA; tmsjohnston@ucsd.edu
* Correspondence: crrfpalau@gmail.com

Received: 22 July 2020; Accepted: 11 August 2020; Published: 4 September 2020

Abstract: Sea surface temperature, determined remotely by satellite (SSST), measures only the thin "skin" of the ocean but is widely used to quantify the thermal regimes on coral reefs across the globe. In situ measurements of temperature complements global satellite sea surface temperature with more accurate measurements at specific locations/depths on reefs and more detailed data. In 1999, an in situ temperature-monitoring network was started in the Republic of Palau after the 1998 coral bleaching event. Over two decades the network has grown to 70+ stations and 150+ instruments covering a 700 km wide geographic swath of the western Pacific dominated by multiple oceanic currents. The specific instruments used, depths, sampling intervals, precision, and accuracy are considered with two goals: to provide comprehensive general coverage to inform global considerations of temperature patterns/changes and to document the thermal dynamics of many specific habitats found within a highly diverse tropical marine location. Short-term in situ temperature monitoring may not capture broad patterns, particularly with regard to El Niño/La Niña cycles that produce extreme differences. Sampling over two decades has documented large T signals often invisible to SSST from (1) internal waves on time scales of minutes to hours, (2) El Niño on time scales of weeks to years, and (3) decadal-scale trends of +0.2 °C per decade. Network data have been used to create a regression model with SSST and sea surface height (SSH) capable of predicting depth-varying thermal stress. The large temporal, horizontal, and vertical variability noted by the network has further implications for thermal stress on the reef. There is a dearth of definitive thermal information for most coral reef habitats, which undermines the ability to interpret biological events from the most basic physical perspective.

Keywords: water temperature; coral reefs; internal waves; upwelling; advection; climate change; coral bleaching

---

1. Introduction

Water temperature is the most easily quantified physical parameter for shallow-water tropical marine environments. Satellite-derived sea surface temperature (SSST) is used by coral reef scientists for global perspectives on temperature (hereafter abbreviated as "T") patterns/trends and coral bleaching status/predictions [1–5]. There is an "implicit assumption" [6] regarding SSST data that "surface temperature metrics provide useful environmental information with respect to corals that typically live meters to tens of meters below the surface". Satellite data have been thoroughly accepted to document patterns of sea surface T (SST) [3,4,7] to such an extent that some studies concerning coral bleaching do not employ any in situ data, relying solely on SSST [8–12]. The 1998 La Niña warmed the waters around Palau and the western Pacific, which caused coral bleaching, high mortality, and the degradation of reefs [13]. In response to this event, a network of in situ T measurements was initiated in Palau to understand SST variability and its effects on reefs (Figure 1). While SSST describes the broad spatial and temporal patterns, this network reveals large amplitude changes, on periods of minutes to years on the outer reef slopes, some of which are invisible to SSST. The overall Palau network presently comprises over 70 stations and 150 compact T loggers (TL), which are deployed by divers [14]. This paper mainly offers guidance from experience on how to set up and maintain an extensive "value-added" T monitoring network in a coral reef area.

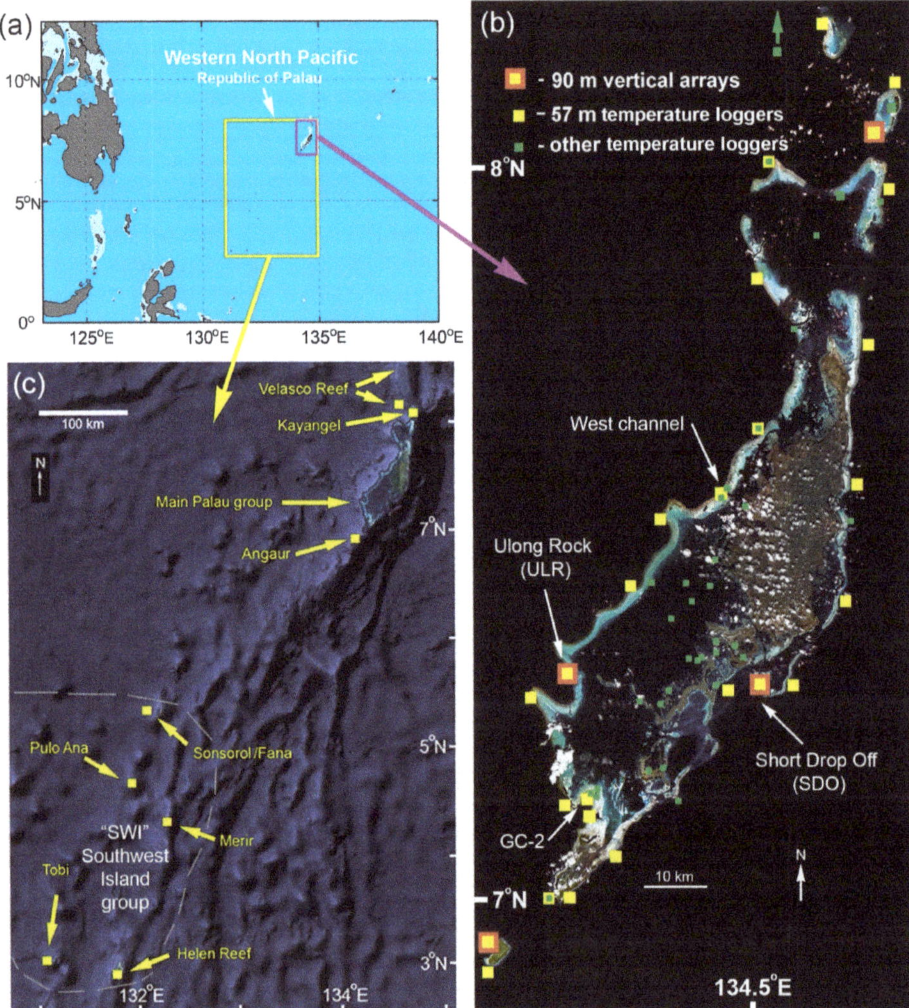

Figure 1. (**a**) General map of the western North Pacific with the location of Palau indicated. The areas shown in (**b**) and (**c**) are indicated by purple and yellow boxes respectively. (**b**) Locations of main temperature monitoring arrays and stations (Ulong Rock, Short Drop Off, GC-2 German Channel, Western Channel) on the outer slope and lagoon regions of Palau. (**c**) Locations of temperature logger stations in outlying areas of Palau, including the Southwest Islands group (SWI) and northern Ngeruangel/Velasco Reef area.

Three concerns about SSST versus in situ data belies uncritical acceptance of the implicit assumption: (1) accuracy compared to shallow (less than 5 to 10 m depth) in situ conditions; (2) diurnal T changes occurring in the upper few meters of the water column; (3) vertical and horizontal variation in thermal conditions. SSST is determined via emitted infrared radiation from the uppermost 20 µm of the water column ("SSTskin") with a 1 to 4 km horizontal resolution and microwave radiation from the upper 1 mm sub-skin ("SSTsubskin") with a circa 25 km resolution [15]. T immediately below these thin layers ("near-surface SST") is not measured by satellite but comes principally from drifting buoys (Global Drifter Program, 20 cm probe depth) and shipboard measurements (various shallow depths of 1 m or more) and are then used to validate SSST data [2]. Oceanic moorings with T probes along their length, such as TAO/TRITON (the nearest to Palau is 260 km away) and ARGO profiling floats (which do not regularly measure T above 5 to 10 m depth) do provide some data at coral reef depths, but almost always in the open ocean far distant from reefs [16]).

The diurnal daytime T changes in the upper few meters of the water column are "visible" to satellites, but not captured as SSST. Solar heating warms water over very shallow reefs [14] with minimum daily SSST measured near dawn [17]. Diurnal patterns expose reefs to higher potential heat stress during the day and important heat content information is lost using only nighttime data. Very shallow stratification can occur, particularly with low wind speeds and calm seas, while wind, waves, and tidal currents can increase shallow mixing and advection of heat at other times. At some depth below the surface, usually a few to several meters, the diurnal heating signal is largely lost, and T remains diurnally stable [14], often to depths of tens of meters. In summary, [15] provide a similar framework for discussing SST defining SSTskin (upper 20 µm), SSTsubskin (upper 1 mm), SSTdepth (about 10 m), and SSTfoundation > 10 m). For our purposes, we refer to SSTskin and SST subskin as SSST. SSTdepth has diurnal variation and SSTfoundation has none. To make sense of these data, environmental conditions at the time of acquisition must be taken into account. On and near reefs, these conditions show extreme variability in periods ranging from minutes to decades, as we show below, some of which cannot be detected in SSST. We focus mainly on SSTdepth and SSTfoundation.

Photophilic reefs occur from the shallows (upward limits determined by sea level) to over 100 m depth in clear tropical waters [18]. Detailed measurement/analysis of thermal patterns over depth ranges and time across a reef tract indicate SSST has a limited capacity to capture the dynamics in the environment [19–21]. On outer slopes once below the diurnally variable depths (usually a few to several meters), "foundation" T in most tropical-subtropical reefs is relatively stable for a few tens of meters. Descending further, however, T (and light levels) decreases in the deepest areas of reefs experiencing consistently lower T, all invisible to satellites. Moving horizontally from ocean slope to inner lagoon or channel areas T patterns vary, but inshore areas are generally warmer. Tidal currents advect water from the ocean to a lagoon (or the reverse) moving shallow water masses of different T/salinity across environments. Adjacent landmasses may influence T in lagoons or over fringing reefs. Even the 5 km pixel size of thermal remote sensing [22] still limits the ability to measure skin T in many areas due to their geography.

The vertical distribution of T over different time scales (decades to minutes) can differ considerably from SSST. During La Niña, Palau's reef waters are warmest. In addition, SSST and 2 m depth in situ values are relatively close, but correlations decrease for in situ data at 15 and 35 m [14]. During ENSO neutral conditions, SSST correlations with in situ T decrease at all depths compared to La Niña. During El Niño, the thermocline shoals and cool water is brought closer to the surface. Surface waters remain relatively warm but the correlation of SSTdepth and deeper in situ data are low, while SSST can be several degrees warmer. Near the lower depth limits of photophilic reefs in Palau (roughly 60 to 70 m) differences between SSST and in situ data are even larger (as much as 12 to 15 °C), with high daily variation in T over both short (min to hours) and long periods (weeks to years, ENSO related).

Accepting that SSST is not a panacea for all consideration of T on coral reefs and that the "implicit assumption" may not be universally applicable, increased in situ measurements are needed to understand the scale of differences in thermal conditions on broad reef areas over depth and time. Only accurate in situ reef T data, the "gold standard" [14], can inform researchers and managers regarding how representative are SSST data and to move beyond generalized SSST values to understand that each locality and habitat is different. For most reef areas, a broad reach of monitoring efforts (temporal, vertical, horizontal) is needed to document representative conditions.

2. Materials and Methods

2.1. Palau and It's Monitoring Network

The Republic of Palau is a western Pacific country with extensive and diverse reef systems located at 3° to 8° N (Figure 1). An extensive barrier/fringing reef (300 km in length) surrounds the main island group. Several lagoons total nearly 1000 km^2 in area with many reefs [23] inside the barrier reef. Reefs are found along or close to the shores of islands, producing close interactions between reefs and land masses. Within 50 km of the main Palau reef tract, an oceanic island (Angaur), a small atoll (Kayangel) and a large sunken atoll (Ngeruangel/Velasco Reef-area 400 km^2) occur. Five oceanic islands and one atoll, collectively called the Southwest Islands (SWI) occur 300 to 500 km southwest from the main reef/island complex (Figure 1c). Found at 3° to 5° N, the SWI are usually within the flow of the eastward North Equatorial Counter Current (NECC). The main group, centered on 7° N, lies at the southern edge of the westward North Equatorial Current (NEC) but is impacted by the NECC as it shifts northward seasonally and during ENSO.

Palau is ideal for examining tropical upper ocean T variability over broad to fine spatial and temporal scales. It has a narrow annual shallow-water T range (1.5 to 2.5 °C), but high variation (over 15 °C) in daily/weekly means in the deeper photic zone. It experienced a major coral bleaching event in 1998 [13], a lesser event in 2010 [12], and a series of small localized events in other years [23]. In 1998, excessively warm water caused loss of zooxanthellae (bleaching) in reef corals resulting in high mortality and degradation throughout the depth range of reefs [13]. At that time there was no accurate T data for Palau's reefs, but in the wake of that event efforts to document Palau's ocean T were started. In 1999, an outer reef vertical array of 5 TLs at depths of 2, 15, 35, 57, and 90 m was set up at Short Drop Off (SDO) on the eastern barrier reef (Figure 1b). Concurrently, a few TLs were installed on shallow inshore/lagoon reefs. Initial array results indicated rapid and extreme variations at the 57 and 90 m stations on the profile. In 2000, a second equivalent array was established at "Ulong Rock" (ULR) on the western barrier reef [24], which recorded some of the largest isothermal vertical displacements ever. Since then the network has expanded to over 70 stations with 150 TLs, ranging from all of Palau's SWI to the north reefs (Ngerunagel/Velasco Reef) with broad coverage of habitats and depths within lagoon areas.

To look at deep reef internal waves, a new initiative deployed TLs in 2014 at 57 m depth at 27 stations, which were widely distributed on the outer slopes of the main Palau reef tract, Kayangel Atoll, Angaur Island, and Velasco Reef. Called the "deep network", they record T once a min, each recording 525,600 measurements every regular year. The depth of 57 m was selected due to the high variability during both El Niño and La Niña periods seen in vertical array data and was a depth from which reasonably easy recoveries/deployments were possible by experienced divers. The deeper stations of the four 90 m depth vertical arrays along Palau's main reefs (Figure 1b) also switched to one-min interval data collection in 2013. In 2015, the deep network was extended to the SWI with five 57 m stations (now six), along with shallower loggers at each location. The deep network, as of January 2020, has recorded 110 million high-resolution T measurements while the remainder of the Palau network accounts for about 10 million more.

Do outer reef slope TL data accurately represent T profiles found in adjacent deep ocean? During 2009 to 2019, Spray autonomous glider missions for Office of Naval Research initiatives [25–27] gathered data on T. At the start and end of wide-ranging missions, data were collected within a few km of the reefs (surface to 200 to 500 m depth) for several days. There was close agreement between open ocean glider and outer slope TL data [28,29].

2.2. Selection of Stations

Care in selecting monitoring station sites is important, particularly where geomorphology or oceanographic conditions may influence thermal conditions. The initial outer reef (SDO and ULR) array sites simply added temperature monitoring to ongoing ecological work, but additional outer reefs stations were subsequently positioned to provide geographic/habitat diversity, such as adding stations on the North and South slope to supplement those on East and West. At some level station selection is "logical" but as a network develops, specific sites may be selected solely on the basis of having an interest in knowing of a detailed area, without knowing in advance whether that station will prove interesting (or not). Lagoon and island stations can be located in what are considered "typical" sites where results may be representative of similar habitats. If a truly distinctive habitat is present, stations can be located to delineate the specific conditions found there, which may differ from nearby areas.

Accessibility during most weather conditions and ease in locating them via distinctive surface/shallow features are advantageous. Each site, even with distinctive features, needs its position established by GPS to a resolution of 0.001′ (circa 1.8 m) and recorded, allowing the site to be located if weather is poor and visibility limited. Sites can also be selected based on additional scientific interest, with other long-term studies being undertaken concurrently (e.g., photo transects) without the need for a separate field trip. T records from such study sites are useful when events (coral bleaching, storms) occur and thermal data are relevant. Stations located in broad shallow areas such as barrier reef tops, that may be accessible only at spring high tides, are often disturbed by storm events and may have few distinctive features (i.e., large coral heads) making relocation of TL instruments difficult. Hence, the need for greater care in choosing sites in these areas is implied and using all available methods to ensure that they can be located later for TL exchange.

2.3. TLs and Their Mounting: Considerations Regarding Installation and Deployment of Instruments

Four types of T loggers have been used: (1) Onset Hobo Pro-8; (2) Onset U22 loggers; (3) Seabird Electronics (SBE) 56; (4) RBR Solo loggers, all relatively small and lightweight. Four factors determined the preferred unit for a station: resolution/accuracy, sampling interval, memory capacity, and response time. In the present effort, target accuracy was 0.1 °C, with a higher precision of up to 0.01 °C with the SBE and RBR loggers. The target accuracy was probably a reasonable compromise compared to SSST data, which is also presented as 0.1 °C. Given the many aspects of variation at individual stations for bleaching indices, the existing accuracy is satisfactory especially considering some of the large amplitude signals.

For all TLs it is important to have a weight attached so it has extra negative buoyancy. The Onset U22 is positively buoyant, the RBR Solo near neutral, and the SBE 56 slightly negative. Instruments and attached weights were deployed, either fixed to the bottom or placed unattached (free) onto the bottom. Mounting instruments on a mooring with float and anchor is risky, as the unit may drift away if the mooring anchor fails, and not advised.

2.4. Characteristics of Instruments

1. Onset Hobo-Pro-8 (Onset Computer Corp., Bourne, MA, USA), units (used 1999 to 2006) recorded to 0.01 °C precision but had to be installed in water/pressure proof housings with only their cabled thermistor probe exposed. Other disadvantages were: limited memory (11 months at a 30 min sampling interval); limited battery capacity; the need to disassemble the housing to replace batteries and download data; continuous problems with leaking in the complicated housing/probe system. This caused some unacceptable loss of data in the early years of the program.

2. The Onset Water Temp Pro 2 (U22)(Onset Computer Corp., Bourne, MA, USA), is (2004 to present) pressure-proof to 120 m depth, has easy deployment, recovery, and download with batteries lasting several years over multiple deployments. With a 30-min sampling interval, the memory records for

two and a half years. Disadvantages are positive buoyancy (requires attachment to a weight), slow response time, a special adapter needed for deployment/download, and difficult battery replacement (return to manufacturer). Details are described subsequently. Cost: US $129.

3. The SBE 56 (Seabird Electronics, Bellevue, WA, USA) (2010 to present) has been reliable, but expensive (four to five times the cost of the Onset U22). Advantages are high resolution, quick response time (exposed probe), long battery life, large memory, and simple set-up and download. It has been used for deep-water stations where rapid thermal changes occur, as well as areas with one to three-year recovery/deployment intervals.

4. The RBR Solo (RBR Ltd., Ottawa, ON, Canada) is similar to the SBE 56 in capabilities/cost and is slightly shorter in length. Advantages are comparable to the SBE 56, but some data have been lost due to power (battery) glitches while deployed. This is presently used in "low priority" locations, where loss of data is not as dire, needing short sampling intervals.

5. Other oceanographic instruments (Onset pressure loggers (Onset Computer Corp., Bourne, MA, USA), Acoustic Doppler Current Profilers, wave gauges, conductivity T loggers) have provided additional T data to the program and their data are considered supplementary, not primary. In a few cases, fast logging dedicated T loggers were attached to longer interval sampling ADCPs or others. Direct data comparisons indicate they can be acceptable substitutes for dedicated TLs.

There are other brands and models of TLs available, but none were used in this project.

2.5. Calibration of TLs

New TLs were calibrated in a water bath at 20 to 31 °C before deployment to establish baseline accuracy against a NIST traceable mercury thermometer, as well as cross calibrating each with other new TLs and older units between deployments. Later re-calibration (between deployments) used similar methods. The target accuracy for all measurements was 0.1 °C.

For calibration, all TLs were set to start simultaneously with a 1-min interval. They were bundled together using rubber bands or similar, placed with probes down while immersed in a circulating water bath and equilibrated for 5 min or more prior to measurements. The mercury thermometer, its bulb situated at the level with the thermistors, was read using a magnifier to 0.02–0.03 °C resolution on the minute when TLs logged values. Calibration runs lasted 30 to 120 min at various T in the water bath. After downloading, data were entered into a spreadsheet comparing each TL against all others in the run and the mercury thermometer. A correction factor was calculated for each logger and applied if needed.

2.6. Sampling Intervals

The measurement interval is a compromise between memory size, battery life, and desired temporal resolution. For most stations with depths less than 35 m, a 30 min interval was used. An Onset U22 with a new battery would have adequate battery life for five to six years at that interval, thus although memory-limited, a new unit could be deployed for two successive periods of 2+ years. In a few instances where Onset U-22 TLs needed to be deployed for over two and a half years, a 1-hr interval was selected to extend memory duration [30].

For thermally active stations SBE 56 or RBR TLs sampled at 1-min intervals. An Onset U-22 sampling at this interval would fill its memory in only 30 days, thus such short sampling intervals were only practical with the more expensive SBE and RBR loggers. At stations where only minor variation in T is anticipated, deployment of high-resolution SBE and RBR TLs may be appropriate if there is a need to verify low variation.

Software for TLs (Hoboware 3.7.18, Onset Computer Corp., Bourne, MA, USA; Seaterm V2, Seabird Electronics, Bellevue, WA, USA; Ruskin 2.10.4, RBR Ltd., Ottawa, ON, Canada) allowed for delayed starts and if several TLs are prepared for array deployment, they should be set to start logging at the same time. Data logged prior to deployment or after recovery should eventually be deleted after download and carefully noting times of first/last good samples (the time the unit is at the proper depth, not just in the water) will allow data to be deleted. Often shallow TLs were recovered while descending to retrieve deeper units, and logged low Ts on such excursions, which needed to be deleted afterward.

2.7. Preparation of Instruments for Deployment

Onset U22. If deployed/recovered by a diver they are best mounted on individual 1 to 2 kg weights, such as the lead weights used for SCUBA diving (Figure 2a). A vinyl plastic cap is placed over the wide "download" end of the TL, to protect the clear plastic window, and the remainder of the TL wrapped in plastic or electrical tape to reduce biofouling. The thermistor is inside the housing near the mounting lug of the housing and this area should be exposed to the water to retain response time.

Figure 2. Various TL deployment methods. (**a**) Onset U22 mounted on dive weight with cable ties. (**b**) Deep network (57 m) mount with a tube for SBE or RBR TL, cable tie lock shown. The line shown used was only for lowering the mount to the bottom, then cut and removed. (**c**) Location of U22 temperature logger (TL) on a very shallow reef positioned in a crevice for protection from direct sunlight, and its location indicated by a large cable tie. (**d**) U22 TL on a reef, rarely (if ever) visited by divers, attached by cable tie and openly visible to aid in locating it for recovery. (**e**) Deep network station (T05) on a steep slope at 57 m depth with marker float to assist in locating the instrument. (**f**) T05 station after five years on the bottom, TL mounting tube attached to two concrete blocks. The logger was exchanged each year for a fresh unit. A marker float is on a line with separate weight (a 15 kg shackle) and not attached to the TL mounting.

One way to install the TL on a weight is to drill a 1/4 to 3/8-inch diameter hole in one margin of the weight. The TL, wrapped in tape with plastic cap in place, is then attached using three plastic cable ties (also known as "zip ties"); one going through both the hole drilled in the weight and the mounting lug of the TL, while two others overlapping diagonally are used to secure the TL at the opposite corners of the weight (Figure 2a). If drilling a hole is inconvenient, additional cable ties help to secure the TL to the weight. Cable ties are readily available, inexpensive, and easily cut if needed. It is tempting to simply attach Onset U-22 loggers to the bottom using a cable tie or similar through the mounting lug and then attaching that to something solid on the bottom. The Onset U22, however, is positively buoyant and if the cable tie fails (which occurs surprisingly often with upward force on the latching mechanism), the unit will drift away. If secured to a weight by at least three cable ties, it is improbable for all to fail and the unit to drift away.

SBE 56 and RBR Solo. These cylindrical TLs have a 25 mm outside diameter, fitting inside a 12 inch (30 cm) long piece of 1-inch (25 mm) inside diameter PVC plastic pipe. The bore of the pipe is usually slightly oversized inside (26 mm), a close fit for the TL but still with some clearance. The TLs, a mounting lug with a 5/16″ (8 mm) diameter hole on one end and exposed probe on the other, can be inserted inside the pipe to a level so only the probe end of the TL is exposed (Figure 2b). Holes of 3/8 to 1/2″ diameter (10 to 13 mm) pre-drilled through the PVC pipe at the correct distance allowed a single cable tie to be inserted through the tube walls and mounting lug, then secured back on itself, to lock the TL into the tube. If holes in the tube for the cable tie "lock" are too small (1/4″ or 6 mm diameter), this can make inserting the cable tie at depth difficult. The TL can be exchanged at depth by a diver cutting and removing the single cable tie, pulling the TL out of the tube, inserting a new unit to the correct level, then inserting new cable tie and locking in place. The SBE and RBR TLs are of different lengths and require holes at different distances along the tube (19/26 cm) to have just the probe exposed. When preparing mounting tubes, it is convenient to drill holes at the correct distance for either type of TLs, so tubes are interchangeable.

A completed mounting tube (with or without the TL already installed) can be attached to a 1 to 2 kg weight or another object for securing on the bottom. For the "deep network", standard mountings were made from two concrete blocks tied securely together and with the mounting tube permanently installed on one surface (Figure 2b,f). The blocks and tube are not recovered, instead the diver exchanges the TL by cutting the retaining cable tie, pulling out the TL, inserting a new TL to the proper level in the tube, and then inserting and locking in place with a new cable tie. With practice, this can be accomplished on the bottom in less than 1 min. The recovered TL is placed in a mesh bag carried by the diver and brought up. The time of recovery to the minute should be noted to establish "last good" and "first good" values for the deployment. Mounting instruments on moorings consisting of an anchor weight (or line tied to the reef), line and float is not recommended. They can easily drift away if the mooring fails. In some cases, marker floats are helpful in finding TLs, particularly if diving from a shot line (Figure 2e), but if marker floats are desired, they should be installed separate (unattached) from the TL itself (Figure 2f).

2.8. Divers and Diving Deployments

Nearly all TL deployments were done by free-swimming divers. To depths of 35 m, TLs were readily deployed by recreational-level trained divers. Using nitrox (oxygen-enriched air), instead of compressed air, allows extended bottom times at or quicker repetitive dives to 30 to 35 m depth. Deeper deployments and exchanges were accomplished using deep-air diving (to 60 m) or trimix (helium/air) gases (60 to 95 m depth), requiring specialized training and equipment. Such "technical diving" has become increasingly common, and given the proper equipment, training, and planning, it is feasible to diver deploy/recover TLs to depths of 100 m [31]. In Palau, there were numerous individuals, usually working in the tourism dive industry, who were capable of doing this type of diving; not an unusual situation considering the worldwide development of diving tourism.

2.8.1. Shallow Deployments

"Shallow" TL deployments are considered to range from the surface layer to the depths of 30 to 40 m with no decompression diving. Across this range, different conditions are encountered, and deployments optimized for the conditions expected. A primary consideration when deploying is the ease with which a TL can be located and recovered months to years later. Over time it may become covered in marine growth or the structure to which it was attached crumbled or gone.

TLs secured to a small weight may be put out on the reef, either free on the bottom (not secured, but in a location where their weight makes it unlikely to be moved by currents or waves) or attached to the bottom. The use of SCUBA weights for TLs has the advantage that the belt loops designed into them are convenient openings through which to run cable ties to secure them to the bottom (Figure 2a). Cable ties come in various lengths, making it easy to go around large objects or to thread through small openings for securing TLs (Figure 2d). Often an area of rocky reef will have small rock arches where the TL can be secured by a long cable tie.

In some cases, it is easier not to attach a weight/logger to the bottom. If the unit is located in a slight depression where being moved is unlikely, it can just be dropped off without further attachment. Nevertheless, ideally all TLs should be attached to the bottom, even if only loosely, since if left unattached and it disappears, it is necessary to search around the site (often difficult at depth) and if not found, no reason can truly be ascribed to its loss. The trade-off is that it takes longer to deploy and recover the unit and in deep water seconds are important in this process.

2.8.2. Special Considerations for Very Shallow Sites

Sites less than 5 m deep should be considered "very shallow" and in areas exposed to waves and swell, the TL must be mounted in a manner capable of surviving very rough conditions. TLs securely attached to weights, still need to be attached as a unit to a strong point (rock) on the reef itself. Use of heavy-duty cable ties (strap width 3/8 inches or more, with a breaking strength of 100 kg or more) is the easiest method for securing units in rough areas. Multiple cable ties should be used, and efforts made to ensure the unit once secured cannot be moved, otherwise, over time the unit may work loose and be potentially lost.

2.8.3. Shelter for Loggers

In very shallow water, direct sunlight may heat a logger resulting in values in excess of in situ water Ts with an observed an increase of 1 to 3 °C for loggers (depending on model) submerged in a flow-through mesocosm, but openly exposed to the sun [32]. It is recommended to either place the logger in a location sheltered from direct sunlight or install a reflective shade over the unit.

2.8.4. Marking Sites

Whether or not to hide TLs on the reef is an important consideration. For shallow water deployments in areas frequented by divers or fishers using SCUBA/snorkeling equipment, it is critical to hide loggers so that they are not seen and picked up inadvertently. Once coated in biofouling, a TL often looks like something that has been lost, perhaps dropped from a boat; not something that has been put in place for scientific research. When TLs are hidden, tucked into crevices or beneath overhangs, it is important to be able to find them again. Marking the general site where a TL is located (but not the TL itself) is helpful, with cable ties attached to a rock or projection nearby as an easy solution (Figure 2c). Photographs of the area (Figure 2d) are also very helpful, covering the area nearby, showing marking cable ties or showing locations of features, which are distinctive to help identify locations of TLs.

In areas where someone finding and picking up a TL is unlikely, efforts can be directed instead at making the TLs more visible underwater and easier to find. Attaching a number of long cable ties (30 to 60 cm), either to the weight or in the areas of the TL, so they stick out in various directions are occasionally helpful in relocating TLs that have been deployed for some time (Figure 2c). The straight tails of the cable ties sticking out are distinctive, even if covered in algae. At times, cable tie tails sticking out of sediment from a buried TL has been the only portion of a TL visible and allowed recovery, otherwise the TL would have been lost.

Covering the TL in colored tape reduces biofouling and helps locate it (yellow is particularly distinctive; Figure 2a). Marking shallow sites with small floats aids visibility at some distance underwater. Careful notes should be taken on the nature of sites, times of deployment, exact depths (using a digital dive computer), and any other features that would help in recovery later.

2.8.5. Photos of Sites

Underwater photos of the sites, particularly showing the exact location where TLs are deployed, are important. Ideally, the TL should be visible in some photos (Figure 2d), to provide clues as to its location after it might be hidden in the reef. Underwater photo trails have proven particularly useful for locating sites on steep reef faces, with a shallow water (5 to 15 m depth) marker (sub-surface float, long cable ties) as a start point. A GoPro or similar camera taking time-lapse photos at 2-sec intervals helps document the routes swum when deploying the unit and can often be retraced if needed all the way to the location of the TL. This technique is especially helpful if the diver recovering the TL did not do the deployment.

2.8.6. Diving for Shallow Reef Deployments

If deployments are needed at depths of 30 to 40 m (the lower limits for "sport diving"), nitrox diving is advantageous as it allows increased bottom time (15 to 20 min) without decompression. If TLs are to be deployed/recovered at various depths moving along transects up a reef slope, the deepest should be done first, then move into shallower water. If doing a number of "swap-outs" of TLs, recovering old ones and setting out new ones, divers should have two mesh "dive bags"; one for recovered TLs and the second for ones to be deployed. The ones being deployed should be labeled indicating depth for each one to avoid confusion while diving. This prevents mistaking a just recovered TL for one that needs to be deployed, which is easy to do in the rush of a working dive.

2.8.7. Deeper Deployments

A number of countries, such as Australia and the United States, have regulations for the workplace and scientific diving which prevent (or make onerous) accessing depths below sport diving limits. Some offer dispensation to undertake advanced diving for research purposes [18]. In Palau, there were no regulations restricting experienced/trained individuals from undertaking the necessary dives to service the TL networks.

Careful selection of areas for deep deployments can simplify the diving involved, particularly for vertical arrays of several instruments. Near-vertical reef profiles are preferred, as this avoids divers having to make time-consuming horizontal or sloping swims between different depth stations. Reef markers, such as very long cable ties or subsurface floats, can be used to mark the route down the slope and individual stations to simplify finding TLs at depth. At the four vertical array sites to 90 m depth (Figure 1b), an experienced diver can exchange all TLs with only a short decompression. Obviously, extra decompression should be done for safety, but the decompression obligation with an efficient dive is not great. It is still incumbent on all persons using these techniques to understand the risks and difficulties involved, prior to attempting what are still very serious dives with risks from decompression sickness (bends), gas toxicity, and drowning. In many cases, scientists may prefer to have diving professionals do deep TL deployments. There is no operational need that would prevent such well-trained competent divers from doing so. As technical diving has become more common, in many locations advanced divers using mixed gas rebreathers can deploy and recover deep TLs.

2.8.8. Shot Line Diving

In some locations, the bottom will be sloping to such an extent that descending and ascending along that slope is not practical as it would require excessive bottom time at depth. In such a situation, the preferred technique is to dive using a "shot-line", an anchor weight (5 to 15 kg) with a line sufficiently long enough to reach the surface from the water depth and a large surface buoy as a position reference. The anchor weight is connected to the line by a clip, which can easily detach the line from the weight, so divers can potentially use the line for ascent leaving the weight on the bottom. When diving on a "shot line" the boat is not anchored ("live boat") but would hold station near the surface float during the dive. Assuming an accurate GPS position is known for the site, the boat slowly approaches the position and once directly over it, the shot line anchor is dropped, and the line is allowed to run free as the weight descends vertically to the bottom. Once on the bottom the remainder of the line and the float(s) are thrown overboard, so the shot line is free of the boat. Ideally, this should be done in conjunction with a depth sounder in the boat, to verify that the sounder depth agrees with the known depth of the TL. Once established on the bottom, the shot line float will indicate if there are currents by leaving a wake behind it, and if excessive, they may drag the anchor over the bottom. The location of the buoy should be monitored for several mins to make sure it is not being dragged over the bottom by currents (which would mean it is no longer at the location of the TL), then the divers prepare and dive.

The divers enter the water at the surface float and swim downward in constant sight of the line. Once near the bottom, they look for the TL location, which should be no more than a few meters away. If a subsurface buoy marking the TL location has been installed earlier (Figure 2e), it may be visible some distance above the bottom and quickly guide the divers to the TL. If not immediately visible, divers search along the correct depth for the TL (assuming a slope).

After locating the TL and exchanging instruments, the divers ascend along the shot line. The weight is detached and either abandoned or sent to the surface with a lift bag. The divers then use the shot line for the ascent. The boat remains close by and once the shot line weight with lift bag comes to the surface, it is picked up and then the boat follows the divers via the surface float. Bubbles should be visible on the surface as well. Divers complete any decompression hanging on the line in mid-water and after surfacing, are picked up by the boat. This technique is efficient when properly used, but like all deep-diving methods, divers need to be carefully trained and completely comfortable with the techniques involved.

In situations where diving cannot be used for deep TL deployment and recovery, remotely operated vehicles [33] or submersibles [34] could be used. These represent a major escalation in costs with uncertain recovery prospects, making extremely difficult what is often a relatively easy process if done by properly trained divers using advanced diving techniques [31].

2.8.9. Deep Moorings

It is tempting to deploy a heavy anchor weight, with a TL near the bottom on a long float line (20 to 30 m for a 57 m deployment). Often such floats are lost and if attached to the anchor weight create extra drag via the mooring line that may pull the entire mooring some distance. Such losses occurred with a few of the original deep network deployments.

One exception to this generality was an extremely deep difficult deployment at Ngaraard Pinnacle [30]. Due to the depth (95 m) over a 100 m deep bottom, currents, and the need to ascend in open water, the TL had been installed on a short mooring line with a weight and depth resistant plastic float and dropped from a boat. To recover the instrument two years later the mooring line was cut by a diver below the TL, allowing the float to carry the TL to the surface (where it was recovered by a boat) abandoning the weight on the bottom. A replacement TL mooring was dropped at the same spot, again by boat, with eventual plans to recover it in the same manner.

2.8.10. Data Management and Availability

Loggers were downloaded using the relevant software and raw data files archived on multiple computers. The raw data, which covered variable periods of time, were saved as Ascii text (i.e., as comma-separated values) or Excel files for further use. The variable deployment length data were put into discrete annual spreadsheets for each station/depth at the appropriate time resolution. The spreadsheet calculated daily/weekly means, range, and standard deviation.

Data are not presently in an open-access database, as they are actively being worked upon prior to publication. Selected data are made available to researchers on request via the CRRF website, which has a catalog of data files (https://wtc.coralreefpalau.org/).

3. Results

Data from TL networks can be used to address general questions (T regime of a reef area) or focus on specific patterns or dynamics within a larger reef tract (e.g., coral bleaching and ENSO changes). Here we provide (1) analysis of thermal patterns and variance on the Palau reef tract; (2) examples of why such a network is useful; (3) how it can be focused on obtaining information on poorly known aspects of reef science. Some examples shown use data from the 2015 to 2016 El Niño, as it was an exceptionally dynamic period, but other years (such as 2010 with coral bleaching) would have been equally informative.

3.1. Outer Reef Temperatures Over Time and Depth of "Coral Reef Conditions"

A vertical array of TLs (i.e., at several depths along the outer reef slope; Figure 3a) shows that the vertical profile of T is an important determinant of maximum reef depth. All TLs recorded at the same interval with an appropriate temporal resolution to document change over time periods ranging from minutes to years (Figure 3b) [14]. The initial array deployments (Figure 1b) on the East and West reef slopes (Short Drop Off in 1999 and Ulong Rock in 2000, Figure 3c) quickly identified two variable temporal patterns on the lower slope. Long-term (month/years) patterns of T varied by about 13 °C, while short term variability (min/hours) was almost as large at about 8 °C (Figure 3d). Initially, a 30-min sampling interval was used due to memory and battery limitations. After 2014 with new TLs, intervals were shortened to 1-min providing evidence of significant variation beyond that measured with a 30-min interval. The deepest depth of the array at 90 m (due to reef slope morphology and diving limitations) proved definitive because T values at 90 m were often well below the accepted limits for coral reefs. Based on data from all depths, 60 to 70 m was established as the approximate lower depth limit of coral reefs is Palau [29].

If full 30-min or 1-min data (respectively 17,560 and 525,600 data points per regular year) for multiple depth TLs at a single station are plotted together, patterns of short and long-term changes over that year are often evident (Figure 3c). The 1-min interval, in particular, reveals new dynamic patterns, such as during the El Niño of 2015 to 2016 [35]. Early in 2016, Ts at 57 m depth were much cooler than at 15 m depth, with a difference of roughly 10 to 12 °C, indicating a highly stratified water column. In March 2016, Ts at 57 and 90 m depth started increasing, along with a lesser increase at 11 m depth, rapidly hitting peaks from June to July 2016 (Figure 3b). The shift over 10 weeks away from strong El Niño conditions produced a "quasi" coral bleaching event [14], which reversed in July with shifts in oceanic conditions [35].

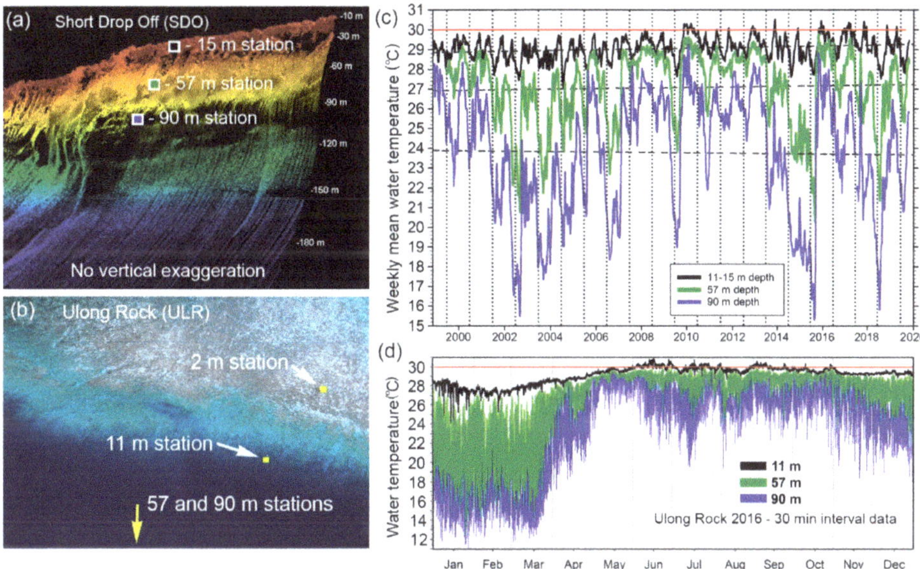

Figure 3. (**a**) Outer slope profile showing locations of TLs along the slope at Short Drop Off (SDO). The location of SDO is shown in Figure 1b. (**b**) Aerial view of Ulong Rock (ULR) station with general locations of TLs indicated, showing the horizontal displacement sometimes necessary to position TLs at certain depths. The location of ULR is shown in Figure 1b. (**c**) Weekly mean temperatures on outer reef slopes 1999 to 2020 in Palau. Only three nominal depths (15, 57, and 90 m) are shown; data also collected at additional depths. Nominal coral bleaching threshold of 30 °C is shown by the red line. Dashed lines are straight line regressions of all data at the three depths. (**d**) Year pattern for 2016 showing all 30-min interval data at the same depths as (**c**). Nominal 30 °C bleaching threshold is shown by the red line.

During La Niña, the deep reefs of Palau become extremely warm, with bleaching level T throughout the water column where reefs occur [14]. Mesophotic reef bleaching is largely unknown due to a lack of deep reef T data and surveys of bleaching at depth during times when bleaching conditions are present [18]. SSST provides no information to estimate deep bleaching. The linking of sea surface height (SSH) and SSST has the potential to estimate thermal conditions on deeper reefs [36,37], although this has not yet been incorporated into global bleaching estimations.

3.2. Impact of Internal Waves on Temperature Dynamics

Preliminary work documented the variable nature of deep T in Palau caused by internal waves/tides, suggesting 60 m depths were near the lower limits of photophilic reefs [24]. In 2014, the enhanced "deep network" was set up with 27 TLs recording at 1 min intervals at 57 m depth (Figure 1b), a depth with high thermal variation (Figures 1b and 3c,d). While this number of stations may seem excessive, preliminary data revealed that each station has a distinct short-term pattern of T variation but when all stations are considered together general patterns are apparent (e.g., a coherent diurnal internal wave in Figure 4). While a complete analysis has not been done, preliminary results indicate island-trapped internal waves can circulate part of the way around the outer slope [38].

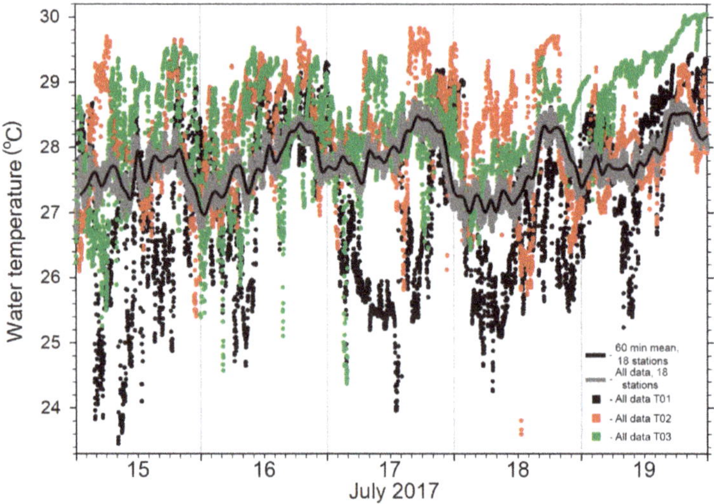

Figure 4. The "deep network" TLs at 57 m depth all around Palau measure once per min providing a wealth of data. This example shows over five days in July 2017, with the hourly mean values of 18 stations (those on the outer slope of the main group) (black line), the mean value each min for those stations (gray area), and examples of raw minute-by-minute data from three stations (T01-black, T02-red, T03-green) in the northern part of Palau indicating the high short-term variation at individual stations.

While most deep network stations are along steep (45° or more) slopes, a few are in areas with lower and more consistent slopes from deep water, where shoaling internal tides (internal waves forced by the semidiurnal or diurnal surface tides) are transformed into internal bores. A station at the South end of Angaur Island (Figure 5) had a remarkable T drop of 12.25 °C in one min (14.4 °C in 3 min). The generation and propagation of these waves are sensitive to stratification [39], which shifted dramatically during the end of the 2015 to 2016 El Niño [35].

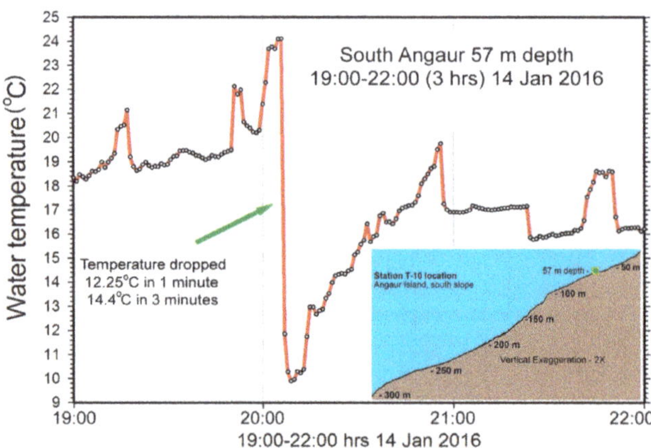

Figure 5. Extreme temperature changes at the southern end of Angaur Island, Station T10 at 57 m depth over three hours. The temperature dropped over 14 °C in three mins, although before and after this event, relatively normal variation in temperature at the station occurred. The gentle slope offshore of the South Angaur station is unusual for Palau's outer reefs (insert-lower right).

3.3. *Mixing of Ocean and Lagoon Water while Advecting through Barrier Reef Channels*

The deep channels bisecting the barrier reef between ocean and lagoon are likely sufficiently deep (35 to 75 m) to ingress water from ocean to lagoon that is vertically stratified, particularly during El Niño periods (Figure 6). Does this occur and how quickly would thermal stratification dissipate as water is mixed and moves into the lagoon? In 2010 a series of six stations with vertical TL arrays at 15, 30, 45, and 57 m (stations five to six without 57 m) depth were set up along the sides of the west/inner channel corridor west of Babeldaob Island (Figure 6a). At times thermally stratified water brought into the channel mouth penetrated several km into the lagoon on a diurnal cycle, but stratification vanished farther into the lagoon (Figure 6c) and water exiting the lagoon on falling tides was well mixed. In August 2010, a La Niña period, coral bleaching was occurring [12] and to a limited extent channels intermittently brought some cooler water into lagoons possibly due to a shoaling diurnal internal tide, which turns into a bore with steep/shallow isothermal slopes on the leading/trailing edge (Figure 6c). The potential effects of these processes on reefs are uncertain.

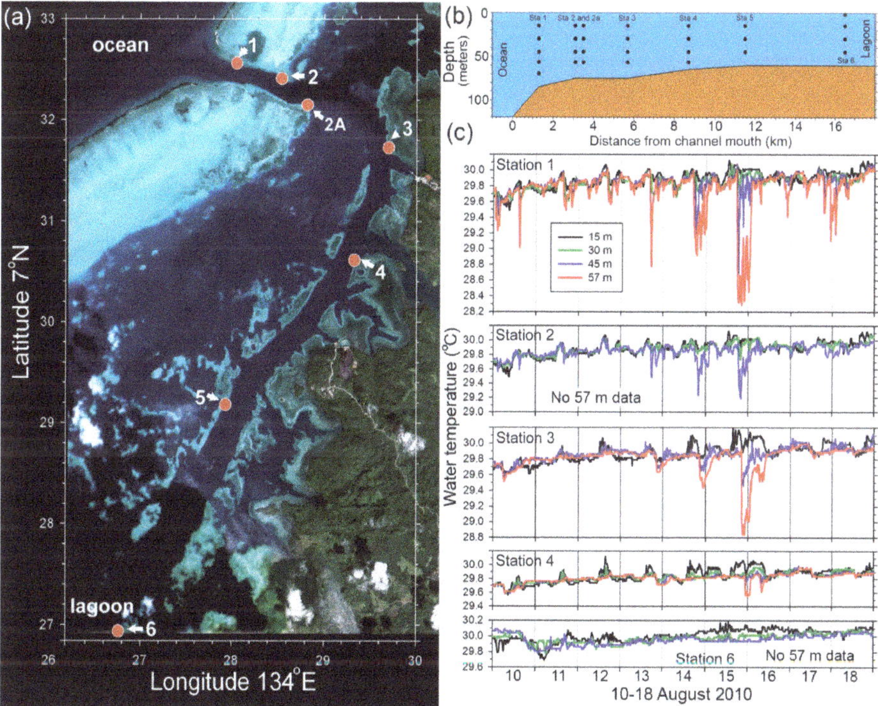

Figure 6. West Channel TL array. (**a**) Locations of stations with vertical arrays of TLs (indicated by numbers) along the sides of the West and Inner Channel into the lagoon. Minutes of latitude (7° N) and longitude (134° E) are indicated on the y- and x- axes. (**b**) Schematic of relative locations of numbered sampling stations along channel axis with the bottom depth of the channel shown. Black dots indicate depths of TLs. (**c**) Vertical structure of water temperature over nine days at stations along the West and Inner channels. Downward spikes in temperature are seen from 15 to 16 August 2010, with stations nearest the channel showing the largest decreases.

Different ENSO conditions change the nature of the offshore water column [28] and are reflected in the water brought into channels on rising tides. During the 2015 to 2016 El Niño water at 35 m depth on the slope of "German Channel" (GC-2 in Figure 1b) had quite variable Ts (Figure 7a). One year later, when the El Niño had dissipated, there was almost no variation in the Ts of incoming and outgoing water (Figure 7b) and again the impacts of this variable T on channel reefs are unknown. Substantial variability is noted around Palau in observations from this network and elsewhere [40]. Furthermore, since these bores are turbulent [41] they can suspend and transport sediment, nutrients, or other properties into shallower water [42,43], which we have noted at this site or nearby.

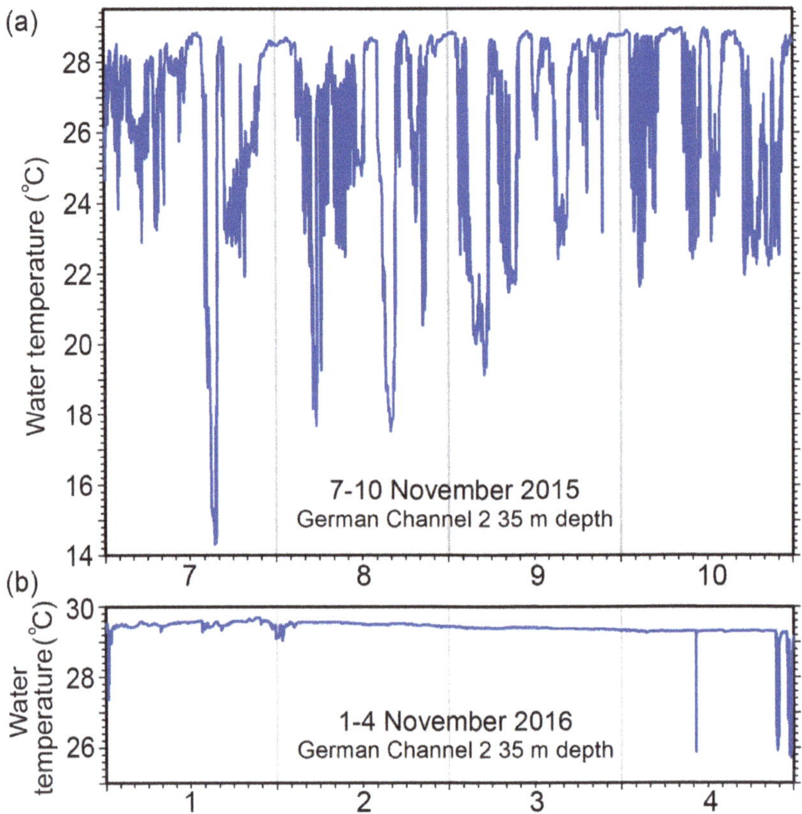

Figure 7. Temperatures at 35 m depth over four days on the slope of the German Channel station 2 (GC-2 location shown in Figure 1b) during (**a**) an El Niño period (7 to 10 November 2015) and (**b**) a period after the end of the El Niño (1 to 4 November 2016), one year later.

3.4. Temperature Patterns across Broad Regions and Current Patterns

With TL stations distributed over a wide geographic range (100 s of km), the network may capture differences attributable to broad oceanographic conditions. The SWI (Figure 1c) are within the eastward NECC, while the main island/reef group is normally dominated by the westward NEC, with occasional intrusions of the NECC. The main Palau group underwent a dramatic shift in currents, sea level, and T in 2016, while the SWI had a lesser shift in thermocline structure at the same time [35]. During the peak of the 2015 to 2016 El Niño in early 2016, shallow (11 to 15 m) daily mean T was very similar between Tobi and the main Palau group (Ulong Rock), separated by 600 km (Figure 8). However, at 57 m depth conditions were very different with Tobi near 25 °C while Ulong Rock was much cooler at 19 to 23 °C. As the El Niño ended in spring 2016, Ulong Rock had a major increase in T over 10 weeks. However, at Tobi T did not spike similarly, but started rising two months later (May) and more gradually; these large differences explained in terms of the dynamics of equatorial currents and equatorial waves [35]. The result is that as the El Niño terminated, water moved back into the western Pacific, and forced the NECC northward towards the main Palau group.

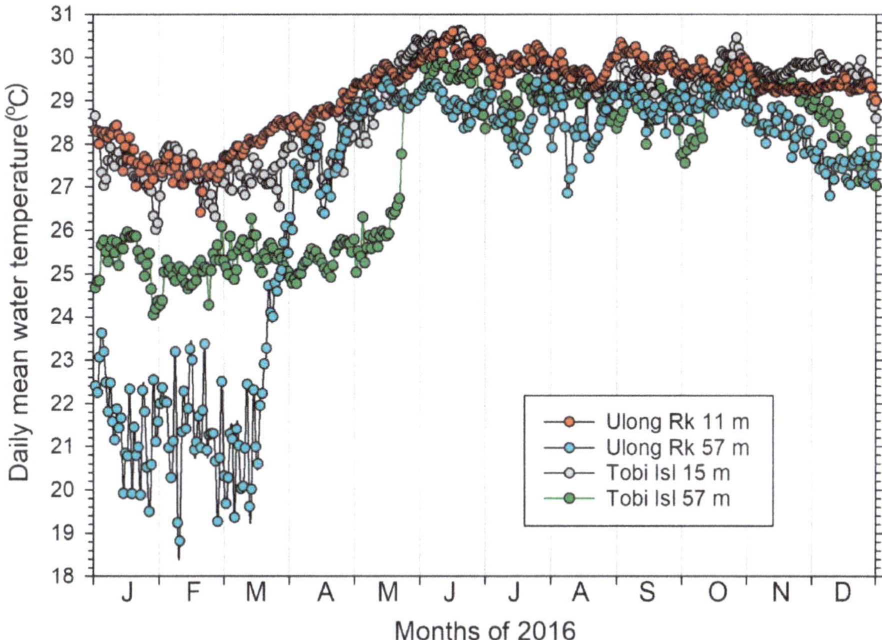

Figure 8. Daily mean temperatures at Tobi, Southwest Islands, and Ulong Rock, main Palau group at 15 and 57 m depths during 2016.

3.5. Climate Change Values from Long-Term Data

To identify climate change (decadal or longer-scale trends) long records are needed, which also resolve considerable variability from internal waves (minutes to hours). ENSO T shifts add another complication, with deeper areas having extreme variation. Periods of months to a few years clearly do not provide a sufficient length to observe whether a climate change signal is present.

With two decades of data from consistent locations, it is now possible to begin examining whether the data show trends potentially related to global climate change. The 11 m weekly mean data (30 min interval) from Ulong Rock ranges from below 27 to over 30 °C (Figure 9) while the 2 m data has a slightly greater range, reflective of diurnal variability in shallower depths. A nominal 30 °C "bleaching threshold" line (Figure 9, red line) [14] indicates this high T level has occurred during several years since 2007, but not between 1999 and 2007. Severe bleaching occurred in 1998, prior to the start of the T network, and temperatures were certainly above the "bleaching threshold". A straight-line regression from the data shows an upward trend of 0.4 °C over twenty years, or 0.2 °C per decade (Figure 9, green line). The rate of increase changes slightly as new data are added each successive year. The trend is about 0.1 °C per decade around Palau from 1971 to 2010 averaged from 0 to 700 m [44], while SST shows a trend of about 0.2 °C per decade from 1900 to 2008 [45]. If our measured trend of 0.2 °C per decade extends to 90 m, it still explains only a fraction (up to 30 mm per decade based on the thermal expansion of seawater) of the 1990s decade long change in sea level in Palau, which appears mainly due to trade wind intensification during that time [46].

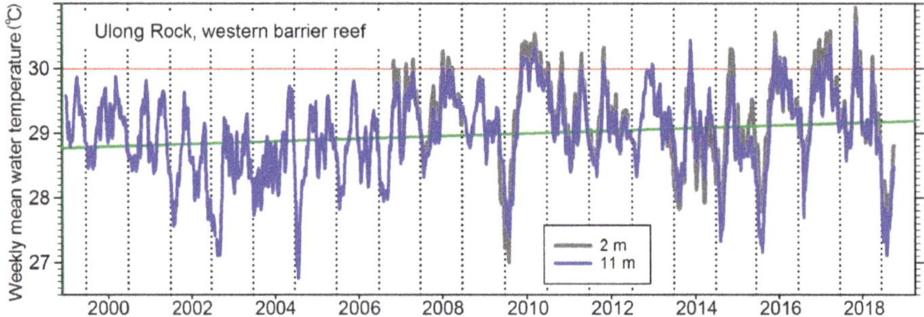

Figure 9. Weekly mean temperatures, 2 and 11 m, at Ulong Rock 2000 to 2019. The red line indicates a nominal 30 °C bleaching threshold above which a few weeks of exposure is associated with coral bleaching. The green is a straight-line regression of the 15 m data indicating a rise of about 0.2 °C per decade since 2000.

4. Discussion

4.1. SSST Versus in Situ Data

Both SSST and in situ data sets have their strengths. SSST provides the global perspective, but in situ data supplies most detail and serves as a ground truth for SSST. In terms of the calculation of heat stress and other T related parameters on reefs, accurate in situ are superior (and essential) when looking at a specific reef or site because of considerable horizontal, vertical, and temporal variability, some of which is invisible to SSST. Monthly mean climatology (MMC) based on SSST is used to define threshold Ts for coral bleaching and calculation of Degree Heating Weeks (DHW), an index of heat stress in a reef area to predict bleaching. If MMC and DHW are calculated from in situ data, the values are different, and one of the strengths of in situ data is that these indices can be determined for locations and depths, not just for single generalized location at the skin surface.

Detailed examination of T and coral bleaching has not been included here, although work has revealed without question the presence of thermal conditions inducing bleaching to 90 m depth in Palau. Deep (mesophotic) reef T conditions and subsequent coral bleaching has been largely ignored in the Indo-West Pacific [18]. Some areas, such as Australia, have experienced severe shallow bleaching events, particularly related to the 2015 to 16 El Niño, and are beginning to examine deep reef Ts/bleaching below 20 to 25 m [33]. While SSST alone cannot indicate Ts at the lower depths where reefs occur, the merging of sea surface height (SSH) data, either from satellites or tide gauges, with SSST provides a new way to assess heat stress in reef environments [36,38] and opens a remote sensing window into events at the lower limits of reefs. While this method required in situ data for validation, it can be expanded to other locations in the tropical Pacific where we expect the relation between Ts and SSH holds.

4.2. Why Is a Network with Many Instruments Needed in Coral Reef Areas?

The Palau network in the main island group provides about 54,000 discrete location–depth–time measurements per day while the NOAA Reef Watch virtual station data for Palau, https://coralreefwatch. noaa.gov/product/vs/data/palau, provides one daily set of SSST measurements (mean, maximum, and minimum), thus cannot capture the thermal dynamics within a reef area, particularly where ENSO related changes are large, internal wave variability is considerable, and diverse types of habitats are present. Without a network appropriately sized for the area to be covered, unknown aspects of the physical environment, along with the biological implications, will be invisible. Each of the vertical TL arrays (four at 2 to 90 m depth, several others at 2 to 57 m), plus the dozens of the widely distributed single 57 m TLs, have shown different patterns. It is not yet clear what is driving these differences, but the ocean current dynamics, documented recently [27,47] and others in the same volume, as well as smaller-scale effects [29], produce a complex physical environment which continues to reveal new layers of complexity. The passages between island groups, such as between Peleliu/Angaur (Lukes Passage) and Kayangel/Ngeruangel (Velasco Reef) and the northern reef tract of the main group (Euchelel Ngeruangel, Kekerel Euchel) are exceptionally dynamic and influence the reefs in those areas greatly. Wake eddies and internal lee waves at 1-km scales are noted at points and over submarine ridges [48–50]. The seasonal shifts of the NEC and NECC, as well as with ENSO, are exceptionally important, changing the nature of the oceanic environment throughout Palau [35]. The same applies to ENSO cycles in the open ocean directly impacting reefs through impressive shifts in the conditions in the photic zone, poorly documented for western tropical Pacific waters [14,25].

The examples presented are largely concerned with outer reef vertical and temporal changes, but examples from back reefs, lagoon patches, and reefs near island shores could alternately have been used. All have different thermal environments and should be within the full scope of T monitoring. Stations inside lagoon areas should provide broad geographic distributions, moving from offshore to inshore habitats and, where water depths are sufficient, established at different depths to capture vertical stratification. The arrays forming the transect of the West Channel (Figure 6) are one extreme of such lagoon arrays. The Rock Island areas of Palau [23], a series of basins separated by sills, have instruments from very shallow to the maximum depths in basins that have minimal water exchange. These inner reef arrays have been important in documenting small scale bleaching events in 2007, 2016, and 2018 [23] for which T information would otherwise have not been available.

4.3. Will a More Modest Network Suffice?

A network of 100 or more TLs may not be feasible (or necessary) for many reef locations. The Palau network developed gradually, and early results indicated the benefits of expanding the network. A more limited suite of TLs can focus on areas where data are most needed. If knowing outer slope conditions relative to reefs is desirable, depth coverage is more important than geographic coverage. Ideally, the deepest levels of reefs in a given area are instrumented. Once the outer slopes of fringing or barrier reefs are covered, inshore areas are then important to determine whether such areas are thermally distinct and if there is any depth stratification of T.

Atolls provide a simpler system (than Palau) to document, usually having a broad scale lagoon circulation. T regimes might differ on opposite outer slopes of an atoll, and vertical arrays in two areas might be informative. Patch reefs within lagoons are convenient locations to establish vertical arrays from near-surface to maximal lagoon depths. Reef flats would also be important to instrument, as they may have significantly higher T. Channels through the reef rim are also important locations for monitoring, as they are the only connections between ocean and lagoon of sufficient depth to ingress stratified water from offshore.

Lagoon areas with islands, such as those that occur in Palau, are more complex and often tidal currents course through shallow channels advecting water to new areas that may have warmed over shallow bottoms. Fresh or brackish water may enter lagoons from streams, springs or groundwater flows, and are another area where documentation would be important. In special cases, such as caves, caverns, siphons, and other areas where groundwater intrudes, different T conditions are expected to occur and should be documented.

4.4. Need for Long-Term Measurements

Short-term in situ T monitoring may not accurately capture broad patterns, particularly with regard to El Niño/La Niña cycles that produce extreme differences [51]. Furthermore, measurement intervals must be short enough to resolve energetic internal waves, although, with sufficient averaging, their effects on a record with long sampling intervals can be reduced. Monitoring networks can take advantage of geography while small oceanic islands can serve as "mooring" equivalents for some global climate considerations. Present-day technical diving capability has expanded the depth range accessible for diver deployment of instruments.

5. Conclusions

We have focused on the techniques for developing and maintaining a network of diver-deployed compact, research-quality T loggers for measuring T from a few meters depth on the reef crest to 90 m on the reef slope. This T network targets a wide variety of environments (reef crest, reef slope, reef channel, atolls, lagoons, pinnacles, and headlands), covers an area from 3° N to 8°30′ N impacted by the NEC and NECC, and for periods over 20 years for some stations.

In terms of ocean physics, the network offers a sometimes astounding view of an energetic environment. With sampling over two decades, we have documented large T signals often invisible to SSST from (1) internal waves on time scales of minutes to hours, (2) El Niño on time scales of weeks to years, and (3) decadal-scale trends of +0.2 °C per decade. The latter is a component of variable sea-level rise in the western Pacific, while the other two signals show 14 °C changes over minutes due to internal bores and over weeks during the termination of El Niño and a dramatic blockage of the NECC's usual path. The T network data have been used to create a regression model with SST and SSH capable of predicting depth-varying thermal stress from satellite measurements, which can be tested now at other locations in the tropics. The large temporal, horizontal, and vertical variability noted by the network has further implications for thermal stress on the reef.

In terms of biology, the data points to numerous areas of investigation, although the program was focused initially on obtaining definitive data on the physical environment that could be correlated with events such as coral bleaching. In general, the program has pointed out a dearth of definitive thermal information for most coral reef habitats within Palau and elsewhere, which undermines the ability to interpret biological events from the most basic physical perspective.

Author Contributions: P.L.C. conceived and developed the network, fieldwork, data analysis and archiving, and writing and editing. T.M.S.J. Field worked on oceanography, analysis and interpretation of data, and writing and editing. All authors have read and agreed to the published version of the manuscript.

Funding: Loggers for the network came from a variety of sources: the University of Oregon, University of California San Diego (UCSD) Senate Marine Sciences Grant, Coastal Observing Research and Development Center of Scripps Institution of Oceanography (UCSD), University the Scholars Fund of UCSD, the David and Lucile Packard Foundation, the University of Delaware School of Marine Science and Policy, the Office of Naval Research FLEAT Initiative, and internal funds of the Coral Reef Research Foundation.

Acknowledgments: Fieldwork was aided by Matt Mesubed, Emilio Basilius, Gerda Ucharm, Sharon Patris, Lori J.B. Colin, Steve G. Lindfield, Paul Collins, Travis A. Schramek, Eric J. Terrill, Daniel L. Rudnick, Jennifer A. MacKinnon, Jonathan D. Nash, Mark Moline, and others.

References

1. Minnett, P.J.; Alvera-Azcárate, A.; Chin, T.M.; Corlett, G.K.; Gentemann, C.L.; Karagali, I.; Li, X.; Marsouin, A.; Marullo, S.; Maturi, E.; et al. Half a century of satellite remote sensing of sea-surface temperature. *Remote Sens. Environ.* **2019**, *233*, 111366. [CrossRef]

2. O'Carroll, A.G.; Armstrong, E.M.; Beggs, H.M.; Bouali, M.; Casey, K.S.; Corlett, G.K.; Dash, P.; Donlon, C.J.; Gentemann, C.L.; Hoyer, J.L.; et al. Observational needs of sea surface temperature. *Front. Mar. Sci.* **2019**, *6*, 1–27. [CrossRef]

3. Heron, S.F.; Maynard, J.A.; van Hooidonk, R.; Eakin, C.M. Warming trends and bleaching stress of the world's coral reefs 1985–2012. *Sci. Rep.* **2016**, *6*, 1–12. [CrossRef] [PubMed]

4. Hughes, T.P.; Kerry, J.T.; Álvarez-Noriega, M.; Álvarez-Romero, J.D.; Anderson, K.D.; Baird, A.H.; Babcock, R.C.; Beger, M.; Bellwood, D.R.; Berkelmans, R.; et al. Global warming and recurrent mass bleaching of corals. *Nature* **2018**, *543*, 373–377. [CrossRef]

5. Skirving, W.J.; Heron, S.F.; Marsh, B.L.; Liu, G.; De La Cour, J.L.; Geiger, E.F.; Eakin, C.M. The relentless march of mass coral bleaching: A global perspective of changing heat stress. *Coral Reefs* **2018**, *38*, 547–557. [CrossRef]

6. Hedley, J.D.; Roelfsema, C.M.; Chollett, I.; Harborne, A.R.; Heron, S.F.; Weeks, S.; Skirving, W.J.; Strong, A.E.; Eakin, C.M.; Christensen, T.R.; et al. Remote sensing of coral reefs for monitoring and management: A review. *Remote Sens.* **2016**, *8*, 118. [CrossRef]

7. Mumby, P.J.; Skirving, W.; Strong, A.E.; Hardy, J.T.; LeDrew, E.F.; Hochberg, E.J.; Stumpf, R.P.; David, L.T. Remote sensing of coral reefs and their physical environment. *Mar. Pollut. Bull.* **2004**, *48*, 219–228. [CrossRef]

8. Guest, J.R.; Baird, A.H.; Maynard, J.A.; Muttaqin, E.; Edwards, A.J.; Campbell, S.J.; Yewdall, K.; Affendi, Y.A.; Chou, L.M. Contrasting patterns of coral bleaching susceptibility in 2010 suggest an adaptive response to thermal stress. *PLoS ONE* **2012**, *7*, 1–8. [CrossRef]

9. Kayanne, H. Validation of degree heating weeks as a coral bleaching index in the northwestern Pacific. *Coral Reefs* **2017**, *36*, 63–70. [CrossRef]

10. Maynard, J.A.; Turner, P.J.; Anthony, K.R.; Baird, A.H.; Berkelmans, R.; Eakin, C.M.; Johnson, J.; Marshall, P.A.; Packer, G.R.; Rea, A.; et al. ReefTemp: An interactive monitoring system for coral bleaching using high-resolution SST and improved stress predictors. *Geophys. Res. Lett.* **2008**, *35*, 1–5. [CrossRef]

11. Muir, P.R.; Marshall, P.A.; Abdulla, A.; Aguirre, J.D. Species identityand depth predict bleaching severity inreef-building corals: Shall the deep inherit the reef? *Proc. R. Soc. B* **2017**, *284*, 1–8. [CrossRef] [PubMed]

12. Van Woesik, R.; Houk, P.; Isechal, A.L.; Idechong, J.W.; Victor, S.; Golbuu, Y. Climate-change refugia in the sheltered bays of Palau: Analogs of future reefs. *Ecol. Evol.* **2012**, *2*, 2474–2484. [CrossRef] [PubMed]

13. Bruno, J.; Siddon, C.; Witman, J.; Colin, P.; Toscano, M. El Nino related coral bleaching in Palau, western Caroline Islands. *Coral Reefs* **2001**, *20*, 127–136. [CrossRef]

14. Colin, P.L. Ocean warming and the reefs of Palau. *Oceanography* **2018**, *31*, 126–135. [CrossRef]

15. Donlon, C.J.; Minnett, P.J.; Gentemann, C.; Nightingale, T.J.; Barton, I.J.; Ward, B.; Murray, M.J. Toward improved validation of satellite sea surface skin temperature measurements for climate research. *J. Clim.* **2002**, *15*, 353–369. [CrossRef]

16. Anderson, J.E.; Riser, S.C. Near-surface variability of temperature and salinity in the near-tropical ocean: Observations from profiling floats. *J. Geophys. Res. Oceans* **2014**, *119*, 1–16. [CrossRef]

17. Gille, S.T. Diurnal variability of upper ocean temperatures from microwave satellite measurements and Argo profiles. *J. Geophys. Res.* **2012**, *117*, 1–16. [CrossRef]

18. Loya, Y.; Puglise, K.A.; Bridge, T.C.L. *Mesophotic Coral Ecosystems*; Loya, Y., Puglise, K.A., Bridge, T.C.L., Eds.; Springer: Cham, Germany, 2019.

19. Safaie, A.; Silbiger, N.J.; McClanahan, T.R.; Pawlak, G.; Barshis, D.J.; Hench, J.L.; Rogers, J.S.; Williams, G.J.; Davis, K.A. High frequency temperature variability reduces the risk of coral bleaching. *Nat. Commun.* **2018**, *9*, 1–12. [CrossRef]

20. Smale, D.A.; Wernberg, T. Satellite-derived SST data as a proxy for water temperature in nearshore benthic ecology. *Mar. Ecol. Prog. Ser.* **2009**, *387*, 27–37. [CrossRef]

21. Venegas, R.M.; Oliver, T.; Liu, G.; Heron, S.F.; Clark, S.J.; Pomeroy, N.; Young, C.; Eakin, C.M.; Brainard, R.E. The rarity of depth refugia from coral bleaching heat stress in the western and central Pacific Islands. *Sci. Rep.* **2019**, *9*, 1–12. [CrossRef]

22. Liu, G.; Heron, S.F.; Eakin, C.M.; Muller-Karger, F.E.; Vega-Rodriguez, M.; Guild, L.S.; De La Cour, J.L.; Geiger, E.F.; Skirving, W.J.; Burgess, T.F.; et al. Reef-scale thermal stress monitoring of coral ecosystems: New 5-km global products from NOAA Coral Reef Watch. *Remote Sens.* **2014**, *6*, 11579–11606. [CrossRef]

23. Colin, P.L. *Marine Environments of Palau*; Indo-Pacific Press: San Diego, CA, USA, 2009.

24. Wolanski, E.; Colin, P.L.; Naithani, J.; Deleersnijder, E.; Golbuu, Y. Large amplitude, leaky, island-generated, internal waves around Palau, Micronesia. *Estuar. Coast. Shelf Sci.* **2004**, *60*, 705–716. [CrossRef]

25. Rudnick, D.L. Ocean research enabled by underwater gliders. *Annu. Rev. Mar. Sci.* **2016**, *8*, 519–541. [CrossRef] [PubMed]

26. Schönau, M.C.; Rudnick, D.L. Glider observations of the North Equatorial Current in the western tropical Pacific. *J. Geophys. Res. Oceans* **2015**, *120*, 3586–3605. [CrossRef]

27. Johnston, T.M.S.; Schönau, M.C.; Paluszkiewicz, T.; MacKinnon, J.A.; Arbic, B.K.; Colin, P.L.; Alford, M.H.; Andres, M.; Centurioni, L.; Graber, H.C.; et al. Flow encountering abrupt topography (FLEAT): A multiscale observational and modeling program to understand how topography affects flows in the western North Pacific. *Oceanography* **2019**, *32*, 10–21. [CrossRef]

28. Colin, P.L. Spotlight on the Palau Island group. In *Mesophotic Coral Ecosystems: A Lifeboat for Coral Reefs?* Baker, E.K., Harris, P.T., Eds.; United Nations Environment Programme and GRID: Arendal, Norway, 2016; pp. 31–36.

29. Colin, P.L.; Lindfield, S.J. Palau. In *Mesophotic Coral Ecosystems*; Loya, Y., Puglise, K., Bridge, T.C.M., Eds.; Springer: Cham, Germany, 2019; pp. 285–299.

30. Colin, P.L.; Johnston, T.M.S.; MacKinnon, J.A.; Ou, C.Y.; Rudnick, D.L.; Terrill, E.J.; Lindfield, S.J.; Batchelor, H. Ngaraard Pinnacle, Palau: An undersea "Island" in the flow. *Oceanography* **2019**, *32*, 164–173. [CrossRef]

31. Pyle, R.L. Advanced technical diving. In *Mesophotic Coral Ecosystems*; Loya, Y., Puglise, K., Bridge, T.C.M., Eds.; Springer: Cham, Germany, 2019; pp. 959–972.

32. Bahr, K.D.; Jokiel, P.L.; Rodgers, K.U.S. Influence of solar irradiance on underwater temperature recorded by temperature loggers on coral reefs. *Limnol. Oceanogr. Methode* **2016**, *14*, 338–342. [CrossRef]

33. Frade, P.R.; Bongaerts, P.; Englebert, N.; Rogers, A.; Gonzalez-Rivero, M.; Hoegh-Guldberg, O. Deep reefs of the Great Barrier Reef offer limited thermal refuge during mass coral bleaching. *Nat. Commun.* **2018**, *9*, 1–8. [CrossRef]

34. Kahng, S.; Wagner, D.; Lantz, C.; Vetter, O.; Gove, J.; Merrifield, M. Temperature Related Depth Limits of Warm-Water Corals. In Proceedings of the 12th International Coral Reef Symposium, 9C Ecology of mesophotic coral reefs, Cairns, Australia, 9–13 July 2012.

35. Schönau, M.C.; Wijesekera, H.W.; Teague, W.J.; Colin, P.L.; Gopalakrishnan, G.; Rudnick, D.L.; Cornuelle, B.D.; Hall, Z.R.; Wang, D.W. The end of an El Niño. *Oceanography* **2019**, *32*, 32–45. [CrossRef]

36. Schramek, T.A.; Colin, P.L.; Merrifield, M.A.; Terrill, E.J. Depth-dependent thermal stress around corals in the tropical Pacific Ocean. *Geophys. Res. Lett.* **2018**, *45*, 9739–9747. [CrossRef]

37. Schramek, T.A.; Cornuelle, B.D.; Gopalakrishnan, G.; Colin, P.L.; Rowley, S.J.; Merrifield, M.A.; Terrill, E.J. Tropical western Pacific thermal structure and its relationship to ocean surface variables: A numerical state estimate and forereef temperature records. *Oceanography* **2019**, *32*, 156–163. [CrossRef]

38. Schramek, T.A.; Terrill, E.J.; Colin, P.L.; Cornuelle, B.D. Coastally trapped waves around Palau. *Cont. Shelf Res.* **2019**, *191*, 104002. [CrossRef]

39. Helfrich, K.R.; Melville, W.K. Long nonlinear internal waves. *Annu. Rev. Fluid Mech.* **2006**, *38*, 395–425. [CrossRef]

40. Leichter, J.J.; Deane, G.B.; Stokes, M.D. Spatial and temporal variability of internal wave forcing on a coral reef. *J. Phys. Oceanogr.* **2005**, *35*, 1945–1962. [CrossRef]

41. MacKinnon, J.A.; Gregg, M.C. Mixing on the late-summer New England shelf- solibores, shear, and stratification. *J. Phys. Oceanogr.* **2003**, *33*, 1476–1492. [CrossRef]

42. Pineda, J. Internal tidal bores in the nearshore: Warm-water fronts, seaward gravity currents and the onshore transport of neustonic larvae. *J. Mar. Res.* **1994**, *52*, 427–458. [CrossRef]

43. Reeder, D.B.; Ma, B.B.; Yang, Y.J. Very large subaqueous sand dunes on the upper continental slope in the South China Sea generated by episodic, shoaling deep-water internal solitary waves. *Mar. Geol.* **2011**, *279*, 12–18. [CrossRef]

44. Flato, G.; Marotzke, J.; Abiodun, B.; Braconnot, P.; Chou, S.C.; Collins, W.; Cox, P.; Driouech, F.; Emori, S.; Eyring, V.; et al. Climate change 2013: The physical science basis. Contribution of working group i to the fifth assessment report of the intergovernmental panel on climate change. In *Evaluation of Climate Models*; Stocker, T.F., Qin, D., Plattner, G.-K., Tignor, M., Allen, S.K., Boschung, J., Nauels, A., Xia, Y., Bex, V., Midgley, P.M., Eds.; Cambridge University Press: Cambridge, UK, 2013.

45. Deser, C.; Phillips, A.S.; Alexander, M.A. Twentieth century tropical sea surface temperature trends revisited. *Geophys. Res. Lett.* **2010**, *37*, L10701. [CrossRef]

46. Merrifield, M.A.; Maltrud, M.E. Regional sea level trends due to a Pacific trade wind intensification. *Geophys. Res. Lett.* **2011**, *38*, L21605. [CrossRef]

47. Simmons, H.L.; Powell, B.S.; Merrifield, S.T.; Zedler, S.E.; Colin, P.L. Dynamical downscaling. *Oceanography* **2019**, *32*, 84–91. [CrossRef]

48. Johnston, T.M.S.; MacKinnon, J.A.; Colin, P.L.; Haley, P.J., Jr.; Lermusiaux, P.F.J.; Lucas, A.J.; Merrifield, M.A.; Merrifield, S.T.; Mirabito, C.; Nash, J.D.; et al. Energy and momentum lost to wake eddies and lee waves generated by the North Equatorial Current and tidal flows at Peleliu, Palau. *Oceanography* **2019**, *32*, 110–125. [CrossRef]

49. MacKinnon, J.A.; Alford, M.H.; Voet, G.; Zeiden, K.; Johnston, T.M.S.; Siegelman, M.; Merrifield, S.; Merrifield, M. Eddy wake generation from broadband currents near Palau. *J. Geophys. Res.* **2019**, *124*, 891–894. [CrossRef]
50. Voet, G.; Alford, M.H.; MacKinnon, J.A.; Nash, J.D. Topographic form drag on tides and low-frequency flow: Observations of nonlinear lee waves over a tall submarine ridge near Palau. *J. Phys. Oceanogr.* **2020**, *50*, 1489–1507. [CrossRef]
51. Skirving, W.J.; Heron, S.F.; Steinberg, C.R.; McLean, C.; Parker, B.A.A.; Eakin, C.M.; Heron, M.L.; Strong, A.E.; Arzayus, L.F. *Determining Thermal Capacitance for Protected Area Network Design in Palau*; NOAA Technical Memorandum CRCP 12; NOAA Coral Reef Conservation Program: Silver Spring, MD, USA, 2010; pp. 1–317.

Journal of
*Marine Science
and Engineering*

Article

Comparison of Satellite-Based Sea Surface Temperature to In Situ Observations Surrounding Coral Reefs in La Parguera, Puerto Rico

Andrea M. Gomez [1,2,3,*], Kyle C. McDonald [1,2,3,4], Karsten Shein [5], Stephanie DeVries [6], Roy A. Armstrong [7], William J. Hernandez [2] and Milton Carlo [7]

[1] Ecosystem Science Lab, Department of Earth and Atmospheric Sciences, The City College of New York, 160 Convent Ave, New York, NY 10031, USA; kmcdonald2@ccny.cuny.edu

[2] NOAA CESSRST, City College of New York, New York, NY 10031, USA; william.hernandez@upr.edu

[3] Earth and Environmental Sciences Program, The Graduate Center, City University of New York, New York, NY 10016, USA

[4] Carbon Cycle and Ecosystems Group, Jet Propulsion Laboratory, California Institute of Technology, 4800 Oak Grove Drive, Pasadena, CA 91001, USA

[5] ExplorEIS, Ashville, NC 28801, USA; karsten@exploreis.com

[6] Biology, Geology & Environmental Science, University of Tennessee Chattanooga, Chattanooga, TN 37403, USA; stephanieldevries@gmail.com

[7] Department of Marine Sciences, University of Puerto Rico, Mayagüez, PR 006682, USA; roy.armstrong@upr.edu (R.A.A.); mcarlo@hotmail.com (M.C.)

*   Correspondence: Agomez@gradcenter.cuny.edu

Received: 20 May 2020; Accepted: 18 June 2020; Published: 20 June 2020

Abstract: Coral reefs are among the most biologically diverse ecosystems on Earth. In the last few decades, a combination of stressors has produced significant declines in reef expanse, with declining reef health attributed largely to thermal stresses. We investigated the correspondence between time-series satellite remote sensing-based sea surface temperature (SST) datasets and ocean temperature monitored in situ at depth in coral reefs near La Parguera, Puerto Rico. In situ temperature data were collected for Cayo Enrique and Cayo Mario, San Cristobal, and Margarita Reef. The three satellite-based SST datasets evaluated were NOAA's Coral Reef Watch (CoralTemp), the UK Meteorological Office's Operational SST and Sea Ice Analysis (OSTIA), and NASA's Jet Propulsion Laboratory (G1SST). All three satellite-based SST datasets assessed displayed a strong positive correlation (>0.91) with the in situ temperature measurements. However, all SST datasets underestimated the temperature, compared with the in situ measurements. A linear regression model using the SST datasets as the predictor for the in situ measurements produced an overall offset of ~1 °C for all three SST datasets. These results support the use of all three SST datasets, after offset correction, to represent the temperature regime at the depth of the corals in La Parguera, Puerto Rico.

Keywords: satellite SST; in situ; coral reefs

---

1. Introduction

In the last few decades, a combination of biotic and abiotic stressors has resulted in significant declines in reef expanse and diversity. Climate change and associated increasing ocean temperatures have resulted in heat stress being identified as one of the greatest threats to coral reef health [1–3]. Tropical shallow water reef building corals live near the upper limit of their thermal tolerance, and a temperature change of as little as 1–2 °C can be detrimental to their health [4]. Further, zooxanthellae experience photoinibition as a result of elevated temperature and light exposure, which damage their photosynthetic systems [5]. Coral bleaching by heat stress involves the production of excess reactive toxic oxygen species, which contribute to oxidative damage and lead to metabolic dysfunction, and sometimes the expulsion of the symbiotic zooxanthellae (i.e., bleaching). Depending on the duration of the event, heat stress can ultimately cause coral death [6–8]. In October 2015, the US National Oceanic and Atmospheric Administration (NOAA) declared the third global coral bleaching event was underway. That global event ended in May 2017, but not before affecting more coral reefs worldwide than previously documented bleaching events, and causing record thermal stress in some areas that had never experienced mass bleaching [9]. Unusually warm sea surface temperatures in the Atlantic were also one of the driving factors for the active 2017 Atlantic hurricane season [10], with some of Puerto Rico's coral reefs being extensively physically damaged by the transit of multiple major hurricanes during the season [11].

The local monitoring of coral reefs by snorkeling or scuba diving provides important detailed information regarding reef health at local scales, but resource limitations restrict the coverage and repeatability of such monitoring to a small proportion of coral reefs globally. The ability to utilize remote sensing techniques to survey corals on broader geographic scales is therefore critical for assessing the effects of anthropogenic climate change in remote or inaccessible areas. Efforts to monitor coral reef environmental conditions in near-real-time on broader (e.g., regional or global) scales currently rely on satellites, because extensive in situ surveys can be cost and time-prohibitive [12]. However, in situ observations at the surface of the ocean, as well at the depth of the corals, are needed to evaluate and improve the accuracy of remote sensing datasets, especially in the shallow, near-shore reef zone, where adjacent land and highly variable benthic albedo can introduce bias in satellite-based measurements [13]. With current technology, the health of coral reef ecosystems cannot be directly observed by satellites in Earth orbit, however, satellite-derived sea surface temperature (SST) data can serve as a proxy for predicting where and when heat stress events can lead to coral bleaching [14].

NOAA's Coral Reef Watch (CRW) program has developed a suite of near real-time satellite SST-based products to monitor heat stress on coral reefs worldwide. CRW SST-based products (Versions 2.0 and 3.0) were used extensively to monitor and document the third global coral bleaching event [9]. CRW calculates the thermal stress for each reef location that can lead to coral bleaching, by comparing near real-time SST values with a long-term SST climatology. The SST climatology supporting CRW's current version 3.1 daily global 5 km product suite is derived from a combination of NOAA/National Environmental Satellite, Data, and Information Service (NESDIS) 2002–2012 reprocessed daily global 5 km Geo-Polar Blended Night-only SST Analysis, and the 1985–2002 daily global 5 km nighttime SST reanalysis, produced by the United Kingdom (UK) Metrological Office, on the Operational SST and Sea Ice Analysis (OSTIA) system. CRW's suite of 5 km products includes SST, SST Anomaly, Coral Bleaching HotSpot, Degree Heating Week (DHW), a 7-day maximum Bleaching Alert Area, and a 7-day SST trend [15]. CRW satellite SST datasets consist of night-only temperature measurements, because this helps to reduce the diurnal temperature fluctuation biases that would occur if both day and night measurements were used [16]. Previous research has also found that in the tropics, night-only temperature measurements agreed more favorably with in situ buoys at 1 m depths [17].

While gridded SST satellite products are usually adequate for monitoring offshore and large spatial areas, the same measurements may not be representative for coral reef ecosystems found in shallow coastal waters [18–20]. A number of previous studies comparing other remote sensing SST datasets and in situ temperature measurements at different geographical locations have found significant differences between these measurements [21–23]. Such discrepancies are caused by a combination of factors, including coarse satellite spatial and temporal resolutions, contamination of the satellite footprint by land areas, and the complexity of the environment in coastal zones (e.g., ocean mixing, increased turbidity, dissolved organic compounds from land [24,25]).

Since coral reef ecosystems are found along coastlines and many coral reef managers use CRW's SST-based products to monitor reefs, it is important to understand the accuracy of these CRW SST products in specific locations, as well as to establish that the SST datasets are optimal [26,27]. Currently, there are no case studies examining the applicability of CRW's (Version 3.1) daily global 5 km satellite SST product of in situ temperature measurements at the depth of the corals in La Parguera, Puerto Rico. To help address this challenge and ensure that CRW's SST products are optimally suited for assessing the temperature at the depth of coral reefs, we deployed a network of in situ temperature sensors at varying depths across four coral reefs at La Parguera, Puerto Rico. Data collected from this network were used as a case study to assess correspondence between CRW's 5 km satellite-based SST product [15], the Jet Propulsion Laboratory's (JPL) 1 km SST dataset [28], and the UK Metrological Office's (OSTIA) 5 km SST product [29]. The CRW satellite dataset is daily, global, and nighttime-only at 5 km (0.05° latitude/longitude) resolution. The OSTIA SST and G1SST are both global daily datasets and support comparison with the CRW SST.

The goals of this case study were to:

1. Assess temperature representation at different coral depths and locations in La Parguera, Puerto Rico, by the three satellite-based remote sensing SST datasets.
2. Identify a statistical model that predicts coral depth temperature from the remote sensing SST datasets.

2. Materials and Methods

2.1. Study Area

La Parguera, located in southwestern Puerto Rico, was selected for the in situ temperature logger deployment (Figure 1), in part because these reefs are extensively environmentally and biologically monitored [30–32], and because NOAA CRW's satellite SST products lack validation datasets in this region. According to paleoclimate data obtained from a coral core, this area has experienced a 2 °C increase between 1751 and 2004 [33]. The local average SST is 27.9 °C, with an annual variability of 3.2 °C (derived from daily measurements between 1966 and 2002 [34]). SST near southwest Puerto Rico are influenced by continental freshwater runoff from the Orinoco and Amazon rivers [35]. This area was affected by the third global bleaching event and the active 2017 Atlantic hurricane season [36,37]. However, the damage from bleaching and hurricanes was non-uniform, and minimal in certain locations.

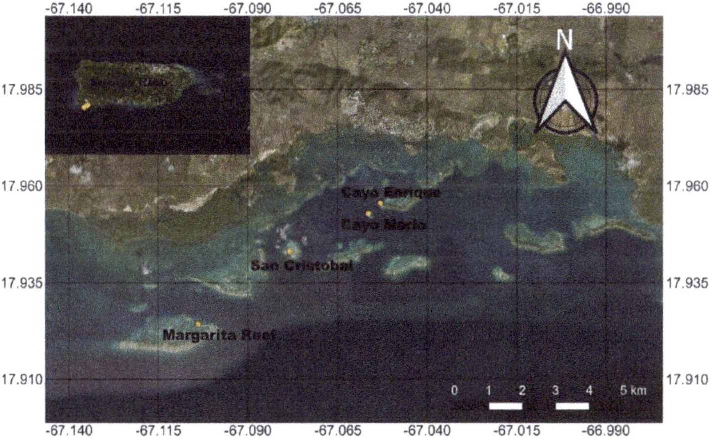

Figure 1. Map of study location, La Parguera, Puerto Rico. During June 2017, six temperature sensors were deployed at Cayo Enrique, and six temperature sensors at Cayo Mario (*n* = 12 total). The depths ranged from 1–13 m. In March 2019, eight temperature sensors were deployed at San Cristobal between depths of 5–6 m, and another eight were deployed at Margarita Reef at depths between 3–4 m (*n* = 16).

The reef system in La Parguera consists of both nearshore and offshore reefs, and is mostly composed of fringing, bank barrier reefs, and submerged patch reefs. The reefs here are experiencing stress from increasing ocean temperatures, rapid coastal development producing an influx of sediments and nutrients, and physical impacts from transitory tropical storms [38].

2.2. In Situ Temperature Loggers

During June 2017, we deployed a total of 12 temperature data loggers in situ within the coral reefs using SCUBA. Six data loggers were installed at varying depths (ranging from 1 m–13 m) at Cayo Enrique (17.95554 N, −67.05312 W), and six loggers at Cayo Mario (17.95283 N, −67.05648 W). The loggers used were HOBO Pendant 64 K Temperature and Light pendant loggers (Onset Computer Corporation, Massachusetts, U.S.A.; ±0.53 °C accuracy at 27.6 °C), and were chosen because HOBO loggers are widely used in coral studies [19,39–44]. These loggers were set to record water temperature at 15-min sampling intervals. Before deployment, the loggers were calibrated against a HOBO Water Temperature Pro v2 Data logger (±.21 °C accuracy at 27.6 °C) for 20 h, to establish relative baseline accuracy and ensure comparability.

The CRW satellite SST dataset was the primary focus of comparison, since it is used by coral mangers to monitor corals. Therefore, in March 2019, an additional 16 in situ temperature loggers were deployed at two new reef locations. This allowed comparison of three, CRW 5 km pixel cells, rather than one; Cayo Enrique and Cayo Mario are situated in one, 5 km pixel. Eight temperature loggers were deployed at San Cristobal (17.94302 N, −67.07834 W) at depths between 5–6 m, and eight temperature loggers at Margarita Reef (17.92422 N, −67.10377 W) at depths around 3–4 m. The temperature loggers used were the same ones deployed at Cayo Enrique and Cayo Mario, with the identical settings. They were calibrated in an ice bath before deployment, to ensure they were functioning within their manufactured accuracy (±0.53 °C accuracy). Table 1 lists information for the depth and location of each logger deployed at all the reef sites.

Table 1. Deployment depth of loggers and associated temperature offset from the calibration.

Location	Logger #	Depth (m)	Offset (°C)
Cayo Enrique	1	10	−0.00759
	2	6	0.081311
	3	5	0.223544
	4	10	0.189653
	5	10	0.151872
	6	6	0.191875
Cayo Mario	7	13	0.195092
	8	11	0.160089
	9	3	0.173979
	10	11	0.137865
	11	3	0.020634
	12	1	0.083416
San Cristobal	1	5	0.343
	2	5	0.232
	3	5	0.343
	4	5	0.232
	5	5	0.232
	6	6	0.232
	7	6	0.232
	8	6	0.343
Margarita Reef	1	3	0.232
	2	3	0.232
	3	3	0.232
	4	4	0.232
	5	4	0.232
	6	4	0.343
	7	4	0.343
	8	4	0.232

The loggers were placed on or near the coral reef framework and secured with zip ties, to minimize movement in the water. Data retrieval was performed in situ, using an optical data shuttle to minimize the potential for observational interruptions, data loss, and instrument disturbances. Night-time observations were extracted for analysis from the time series logger data to compare to CRW's night-time only SST. We evaluated the in situ time series and determined that, owing to the short duration of the study and the use of night-time only observations, drift corrections were unnecessary.

2.3. Satellite Sea Surface Temperature Datasets

The three different gridded, global, daily remote sensing SST datasets listed in Table 2 were assessed in this study. The NOAA CRW program's Version 3.1 products are based on CRW's CoralTemp Version 1.0 SST dataset and are derived from NOAA/NESDIS' operational near-real-time daily global 5 km Geostationary-Polar-orbiting Blended Night-only SST Analysis from late 2016 onwards. OSTIA is also generated globally, and daily, at 5 km resolution, and has been found to be the most suitable SST dataset for aquaculture studies [23]. The SST measurements used in OSTIA are provided by the Group for High Resolution SST, which uses a combination of in situ and satellite data from both infrared and microwave radiometers to generate the SST [29]. The Group for High Resolution Sea Surface Temperature (GHRSST) daily, global 1 km SST (G1SST) dataset is produced by the JPL Regional Ocean Modeling System group. This product is created by using a multi-scale two-dimensional variational blending algorithm on a global 0.0009 degree grid, and is also a combination of data from multiple satellites and in situ data [28]. Remote sensing SST values nearest to the in situ sites were extracted from netCDF4 files by means of a spherical nearest neighbor approach, limited to open water vectors.

Table 2. Remote sensing datasets evaluated in this study.

SST Datasets	Source	Resolution Spatial	Resolution Temporal	Day/Night	References
CoralTemp	NOAA/NESDIS CRW	5 km	Daily	Night	[15]
OSTIA	UKMO	5 km	Daily	Day + Night	[28]
G1SST	JPL ROMS	1 km	Daily	Day + Night	[29]

2.4. In Situ and Satellite SST Statistical Comparisons

To facilitate comparison with the three remote sensing SST datasets, in situ temperature observations between 19:00 and 06:00 (local time sunset/sunrise) were extracted and averaged to produce a daily nighttime-only in situ data series. A Spearman Correlation was performed on the in situ temperature data at each reef site to compare the data from the different logger sites and depths. No statistical difference was found among sites or depth ($p < 0.05$; $r > 0.83$ for all logger sites), indicating a strong correlation between loggers at each site. Therefore, for the statistical computations with the remote sensing SST datasets, the in situ loggers were averaged, to yield one daily, in situ measurement time series. Another reason we decided to average the in situ loggers to produce one daily in situ measurement for each site ($n = 3$), was because some temperature loggers were lost during the course of the study due to hurricanes, earthquakes, and the harsh ocean environment, and of the remaining data loggers, not all were able to capture data by 1 March 2020, when the study ended. On the conclusion of the study in early March 2020, only two loggers were recovered from Cayo Enrique, three from Cayo Mario, three from Margarita Reef, and five from San Cristobal.

For the duration of the study, the daily minimum, maximum, and mean temperatures were computed, along with their standard deviations for Cayo Enrique and Cayo Mario (30 June 2017–1 March 2020), San Cristobal (1 March 2019–1 March 2020), and Margarita Reef (1 March 2019–24 December 2019). The bias (satellite—in situ data) was also calculated for each satellite-based SST dataset, to relate the daily differences between the satellite and in situ temperature logger measurements. A mean bias value above zero corresponds to cooler SST data than the in situ data, and a bias below zero stands for warmer SST data [45]. From the biases, the mean, standard deviation, and root mean square error were calculated. Next, a Spearman's correlation was performed to evaluate the strength of the relationship between the SST datasets and in situ temperature measurements. The correlation coefficient from the test varies between +1 and −1, with values closer to +1, indicating a stronger degree of association between the measurements [46].

Scatter plots of the in situ temperature measurements against the satellite SST were also produced to evaluate the relationship. Finally, a linear regression model was fit to adjust for differences between the remote sensing SST datasets and in situ measurements, with the satellite SST datasets as the predictor for the averaged in situ temperature recorded by the loggers. Given the close, linear agreement between the two variables as observed in the scatter plots, the linear regression model, while not accounting for serial correlation in the residuals, satisfies the goals of this work to fit a model that can effectively estimate reef-depth water temperature from SST estimated by satellite remote sensing.

3. Results

For Cayo Enrique and Cayo Mario, San Cristobal, and Margarita Reef, summaries of the minimum, maximum, mean, and standard deviation (SD) for the averaged in situ temperature and three remote sensing SST datasets for the duration of the study are evaluated, and summarized in Tables 3–5. The mean bias (°C), its SD, and the root mean square error (RMSE) for each remote sensing SST dataset, along with the Spearman correlation coefficient, are summarized for all locations in Tables 6–8 For JPL's Level 4, daily, G1SST dataset, after 9 December 2019 to the present, the SST dataset has been reported to produce poor quality SST data, and this poor data was removed from the analysis (https://podaac.jpl.nasa.gov/announcements/2020-01-29_G1SST_Data_Outage_Alert).

Table 3. Summary of the sea surface temperature (SST) data (°C) for Cayo Enrique and Cayo Mario from 30 June 2017–1 March 2020 (~two years and eight months; n = 975 days) for the averaged in situ loggers (data gap between 11 September 2019–1 October 2019), CoralTemp, and Operational SST and Sea Ice Analysis (OSTIA). The G1SST dataset is from 30 June 2017–8 December 2019 (n = 891 days).

SST Source	Min	Max	Mean	SD
In situ	26.23	30.47	28.41	1.13
CoralTemp	26.21	30.24	28.09	1.02
OSTIA	26.21	30.15	28.15	1.01
G1SST	25.44	30.31	28.21	1.08

Table 4. Summary of the SST data (°C) for San Cristobal from 1 March 2019–1 March 2020 (one year) for the averaged in situ loggers, CoralTemp, and OSTIA. The G1SST dataset is only from 1 March 2019–8 December 2019.

SST Source	Min	Max	Mean	SD
In situ	26.13	30.24	28.37	1.09
CoralTemp	26.28	30.20	28.22	1.03
OSTIA	26.25	30.17	28.25	1.03
G1SST	26.14	30.35	28.57	0.97

Table 5. Summary of the SST data (°C) for Margarita Reef from 1 March 2019–24 December 2019 (~10 months) for the averaged in situ loggers, CoralTemp, and OSTIA. The G1SST dataset is only from 1 March 2019–8 December 2019.

SST Source	Min	Max	Mean	SD
In situ	26.27	30.23	28.55	1.02
CoralTemp	26.27	30.19	28.43	1.02
OSTIA	26.27	30.14	28.48	1.02
G1SST	26.23	30.82	28.55	0.96

Table 6. Mean bias (satellite—in situ), standard deviation of the bias, root mean square error (RMSE) of the bias, and Spearman correlations between satellite-based SST datasets and averaged in situ temperature for Cayo Enrique and Cayo Mario from 30 June 2017–8 December 2019 (about two years and five months; $n = 891$ days).

SST Dataset	Mean Bias (°C)	SD of the Bias	RMSE	Spearman
CoralTemp	−0.35	0.38	0.52	0.94
OSTIA	−0.29	0.37	0.47	0.93
G1SST	−0.33	0.40	0.52	0.92

Table 7. Mean bias (satellite—in situ), standard deviation of the bias, root mean square error (RMSE) of the bias, and Spearman correlations between satellite-based SST datasets and averaged in situ temperature for San Cristobal from 1 March 2019–1 March 2020 (one year) for CoralTemp, and OSTIA datasets. The G1SST dataset is only from 1 March 2019–8 December 2019.

SST Dataset	Mean Bias (°C)	SD of the Bias	RMSE	Spearman
CoralTemp	−0.16	0.33	0.36	0.95
OSTIA	−0.12	0.30	0.33	0.96
G1SST	−0.11	0.31	0.33	0.91

Table 8. Mean bias (satellite—in situ), standard deviation of the bias, root mean square error (RMSE) of the bias, and Spearman correlations between satellite-based SST datasets and averaged in situ temperature for Margarita Reef from 1 March 2019–24 December 2019 (~10 months) for CoralTemp, and OSTIA datasets. The G1SST dataset is only from 1 March 2019–8 December 2019.

SST Dataset	Mean Bias (°C)	SD of the Bias	RMSE	Spearman
CoralTemp	−0.12	0.30	0.32	0.95
OSTIA	−0.07	0.29	0.30	0.95
G1SST	−0.05	0.35	0.36	0.91

All the remote sensing SST datasets evaluated yielded similar temperature patterns, and correspondingly high correlations with the in situ temperature measurements (correlation coefficients >0.91), and the seasonal trends can be observed in the time series (Figures 2–4). A strong seasonal trend is observed, with the three remote sensing datasets consistently underestimating the temperature at the depth of the corals during warmer months (June to September).

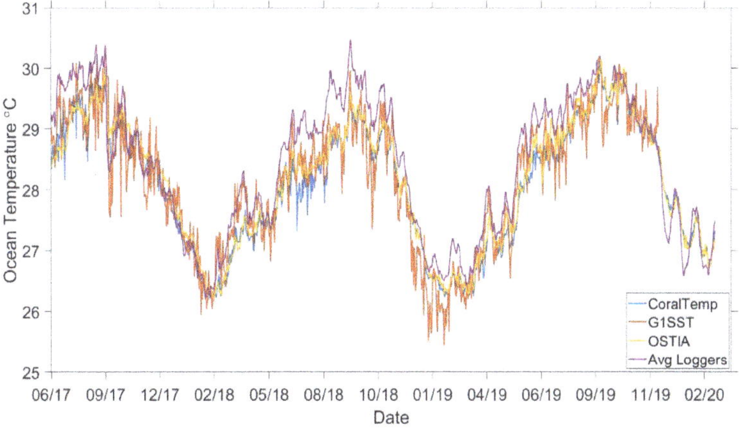

Figure 2. Time series of temperature data from 30 June 2017–1 March 2020 of the three remote sensing datasets (CoralTemp, G1SST, OSTIA) and the averaged in situ loggers for Cayo Enrique and Cayo Mario.

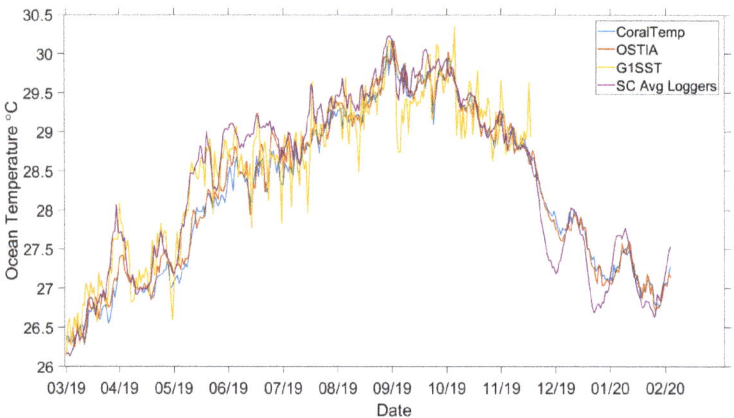

Figure 3. Time series of temperature data from 1 March 2019–1 March 2020 of the three remote sensing datasets (CoralTemp, G1SST, OSTIA), and the averaged in situ loggers for San Cristobal.

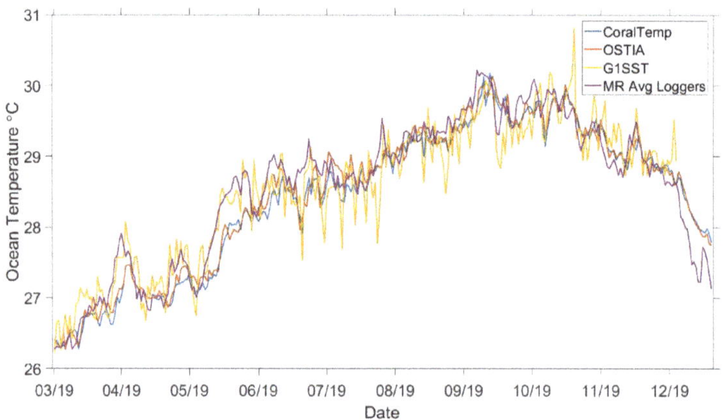

Figure 4. Time series of temperature data from 1 March 2019–24 December 2019 of the three remote sensing datasets (CoralTemp, G1SST, and OSTIA), and the averaged in situ loggers for Margarita Reef.

Ocean temperature seasonality was also explored for Cayo Enrique and Cayo Mario for only two years of the study (30 June 2017–30 June 2019), and it was found that correlation coefficients decreased marginally, but not significantly, when split into the dry and wet seasons. The first dry season (December 2017 to March 2018) had correlation coefficients of 0.90, 0.89, and 0.87 for CoralTemp, G1SST, and OSTIA, respectively. The wet season (April 2018 to November 2018) had slightly higher coefficients of 0.92, 0.89, and 0.95 for CoralTemp, G1SST, and OSTIA, respectively. Overall, all remote sensing SST datasets displayed negative (cool) biases for the majority of the study, and this can be observed in Figures 5–7.

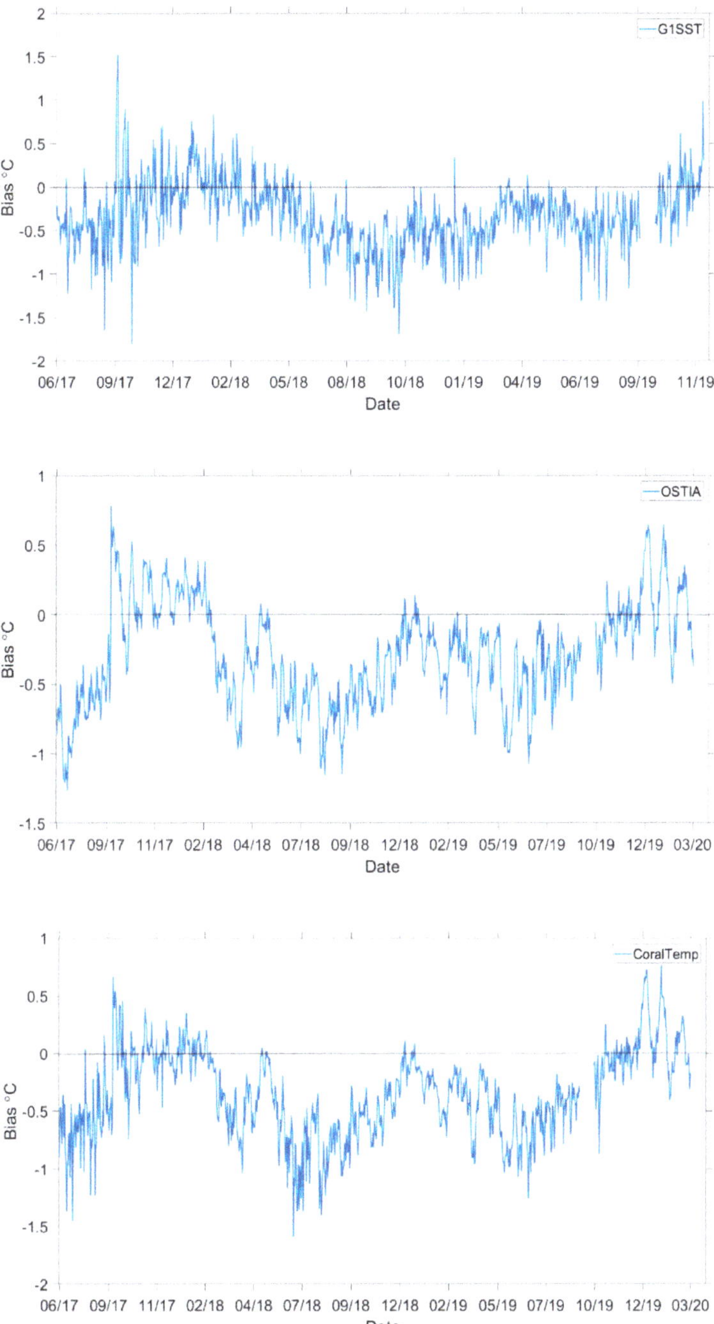

Figure 5. Time series of the bias (satellite—in situ) temperatures from 30 June 2017–1 March 2020 for Cayo Enrique and Cayo Mario.

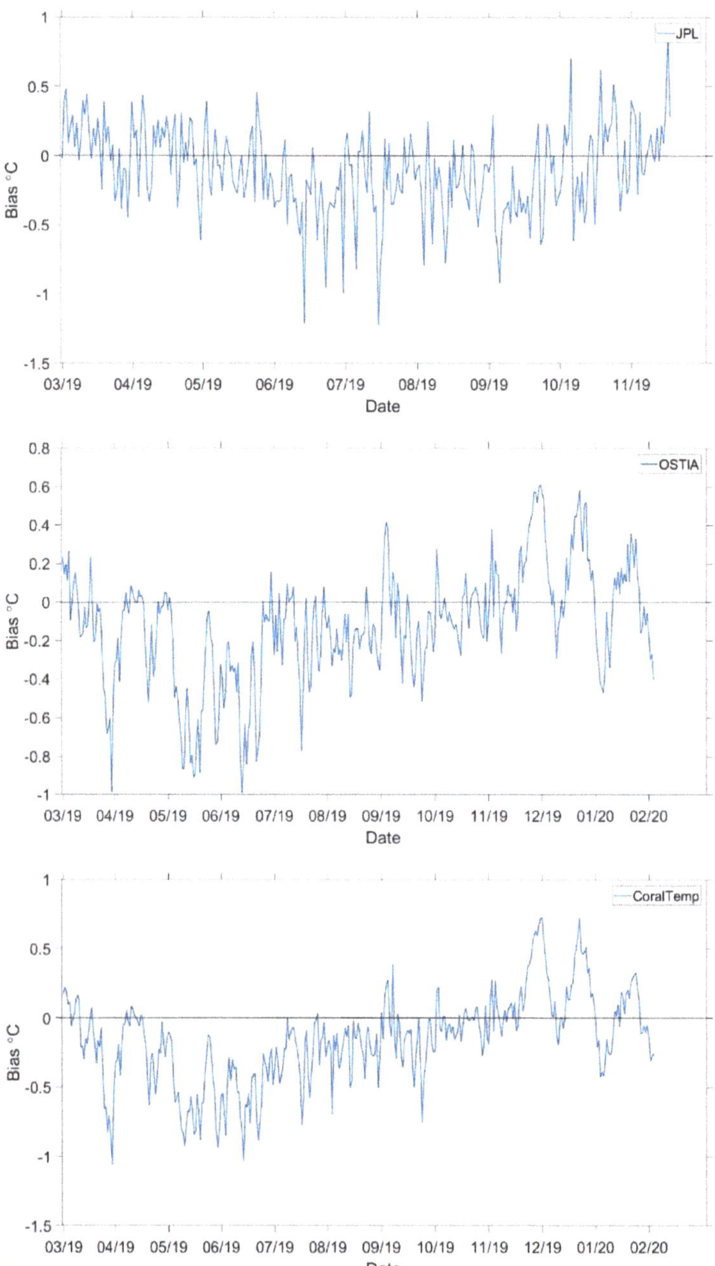

Figure 6. Time series of the bias (satellite—in situ) temperatures from 1 March 2019–1 March 2020 for San Cristobal.

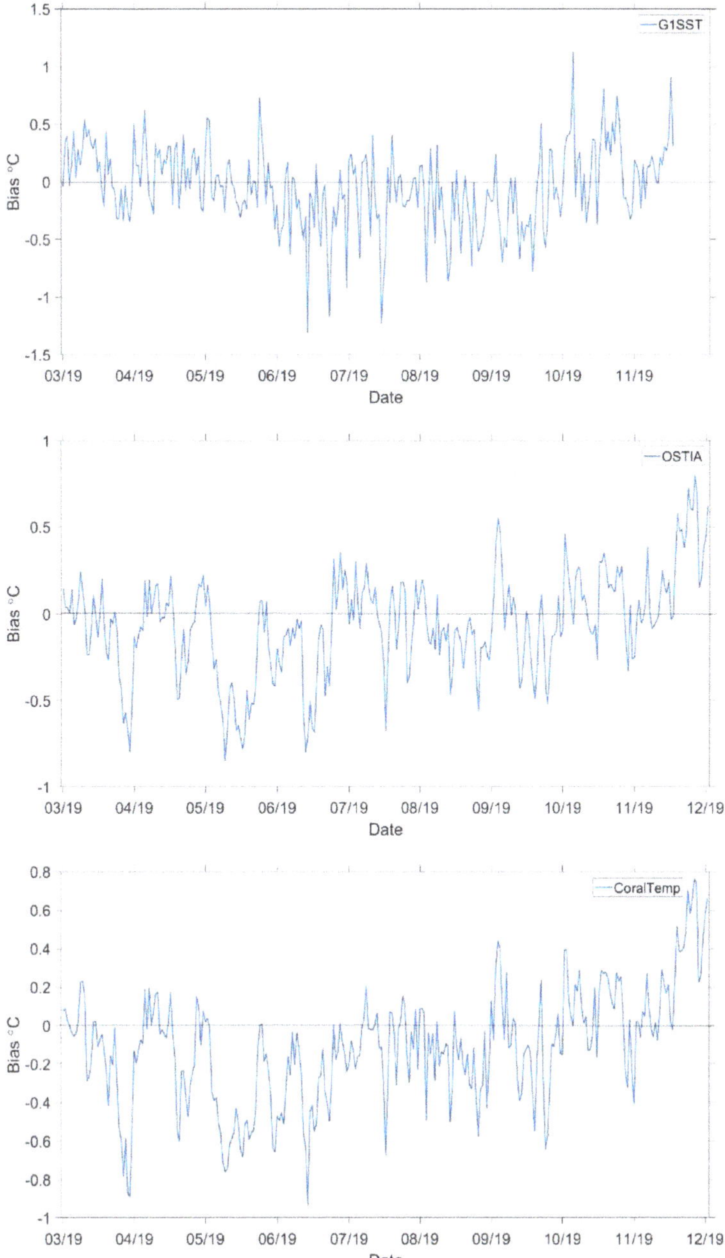

Figure 7. Time series of the bias (satellite—in situ) temperatures from 1 March 2019–24 December 2019 for CoralTemp and OSTIA, and from 1 March 2019–8 December 2019 for G1SST for Margarita Reef.

The scatter plots indicated a strong linear relationship between all remote sensing SST datasets and in situ temperature measurements Figures 8–10. Table 9 contains statistical information from the linear regression models.

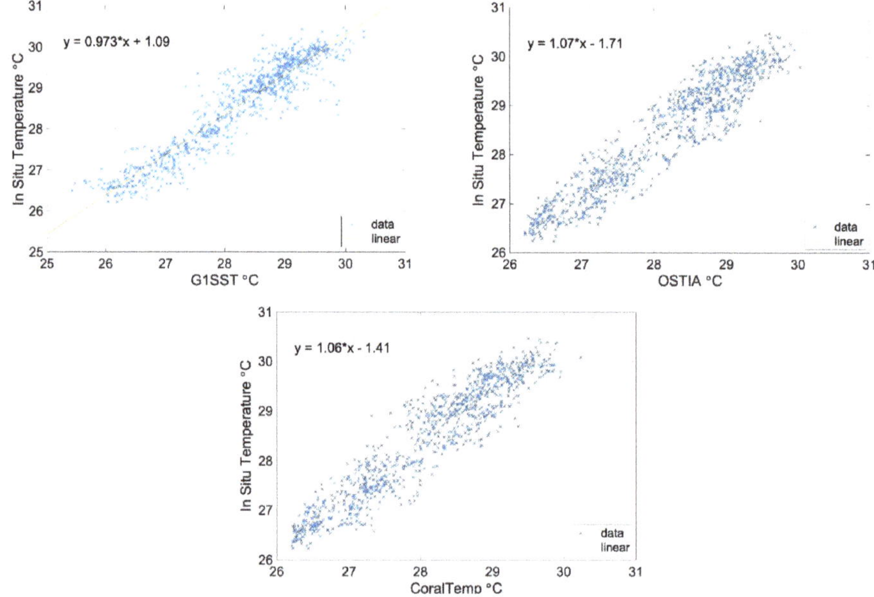

Figure 8. Scatter plots with linear regression of the averaged in situ temperature against the different three different remote sensing SST datasets for Cayo Enrique and Cayo Mario.

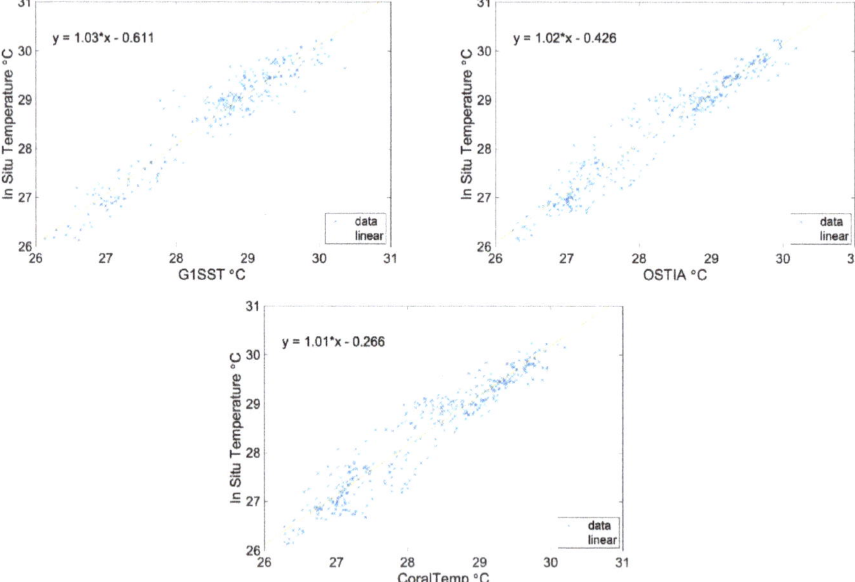

Figure 9. Scatter plots with linear regression of the averaged in situ temperature against the different three different remote sensing SST datasets for San Cristobal.

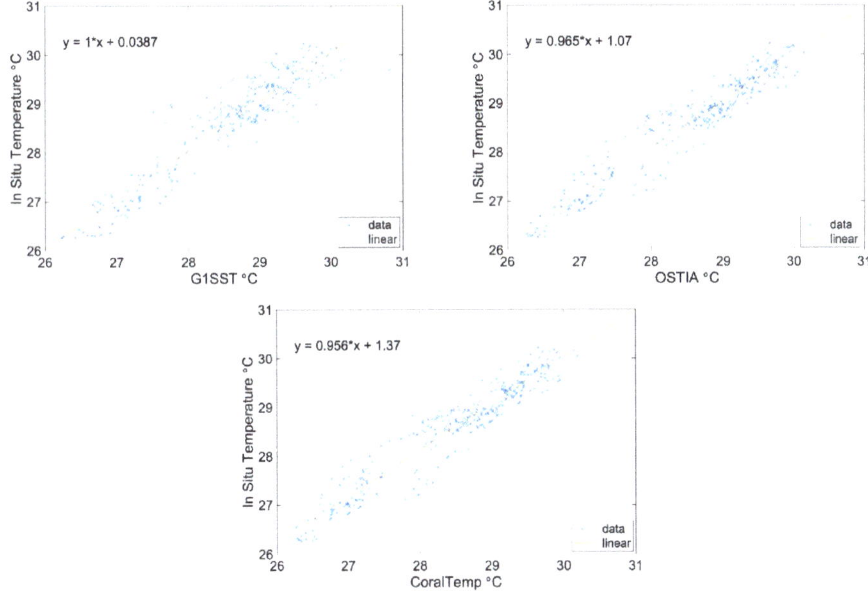

Figure 10. Scatter plots with linear regression of the averaged in situ temperature against the different three different remote sensing SST datasets for Margarita Reef.

Table 9. Test statistics (root mean square error (RMSE); R^2; p-value) from the linear regression models, with the three satellite-based SST datasets (CoralTemp = CT; OSTIA; G1SST) as the predictor for the in situ temperature at the three different sites (Cayo E Enrique and Cayo Mario = CE/CM; San Cristobal = SC; Margarita Reef = MR).

Site	CE/CM			SC			MR		
Statistic	CT	OSTIA	G1SST	CT	OSTIA	G1SST	CT	OSTIA	G1SST
RMSE	0.38	0.37	0.40	0.33	0.30	0.31	0.30	0.29	0.35
R^2	0.89	0.89	0.87	0.91	0.92	0.91	0.92	0.92	0.88
p-value	<0.05	<0.05	<0.05	0.57	0.33	0.27	0.00	0.02	0.95

All remote sensing datasets were negatively biased throughout most of the study period. All scatter plots indicated a strong linear relationship between the satellite-based SST datasets and in situ temperature measurements, suggesting that a linear regression model would be appropriate to estimate coral depth temperatures from the remote sensing data. A linear regression model with CoralTemp, OSTIA, and the G1SST, as the predictor for the in situ temperature measurements yielded an average offset of ~1 °C for all study sites.

4. Discussion

The main goal of this research was to assess the suitability of three remote sensing satellite SST datasets, NOAA's CRW CoralTemp, the UK Met Office's OSTIA, and JPL's G1SST dataset, to capture the temperature at the depth of multiple coral reef ecosystems in La Parguera, Puerto Rico. This study discovered that all three of the remote sensing SST datasets evaluated were acceptable surrogates, after offset correction (~1 °C), of the temperature at the depth in the coral reefs at Cayo Enrique, Cayo Mario, San Cristobal, and Margarita Reef.

Overall, all three remote sensing SST products produced a cool bias (satellite—in situ) within their timeframe of data collection, indicating that the satellite SST was underpredicting the actual temperature at the depth of the corals. For Cayo Enrique and Cayo Mario, the OSTIA SST dataset exhibited the least bias (−0.29 °C, compared to −0.33 °C for G1SST, and −0.35 °C for CoralTemp) for the duration of the study. The G1SST and OSTIA dataset both possessed similar small biases for San Cristobal, at −0.11 °C and −0.12 °C respectively, with CoralTemp yielding −0.16 °C bias for the one-year study. The biases were even smaller for Margarita Reef, with the G1SST bias only −0.05 °C, and the OSTIA SST dataset with a −0.07 °C bias. CoralTemp was found to have a slightly higher cool bias, at −0.12 °C for the ~ten-month study at Margarita Reef. The differences in the satellite SST biases could be attributed to two factors. The first possible explanation is that the study sites were all different depths. The first study sites, Cayo Enrique and Cayo Mario, were overall the deepest sites for the temperature logger deployment (six of the loggers were placed deeper than 6 m, which was the max depth for San Cristobal). San Cristobal was the middle site in terms of depth (5–6 m), with Margarita Reef being the shallowest (3–4 m). The results suggest that the satellite-based SST have a closer temperature matchup with the shallower reefs. This makes sense, because deeper down the water column, the temperature profile drops after the mixed layer (temperature decreases with depth, as it loses availability to sunlight). Another explanation for the bias conflictions between sites is the different time durations of the study for each site. Cayo Enrique and Cayo Mario were the sites with the longest temperature time series (~2 years and 9 months), and they had the relative highest cooler bias offsets for all three satellite SST datasets. Whereas Margarita Reef had the shortest time series (~10 months), and also the smallest biases between the in situ and satellite SST datasets. Longer time series data for all the sites are required to further investigate the biases between the sites and remote sensing SST datasets.

Seasonal patterns were also observed when assessing how representative the satellite-based SST datasets were to the in situ temperature loggers. During the warmer, wet season (June to September), the satellite SSTs often underestimated the temperature at depth recorded by the in situ temperature loggers. This pattern suggests that the satellite SST datasets do not measure the actual temperature at depth when the water is warmer and possibly more stratified [47]. For Cayo Enrique and Cayo Mario, in mid-late August 2017, which is considered to be part of the warm, wet season, a cool bias can be seen in all three remote sensing products. According to NOAA's Climate Review for Puerto Rico 2017, on August 17th and again on the 20th, there was tropical wave activity that created above normal precipitation levels caused by heavy showers and thunderstorms. Then, in September, Hurricanes Irma and Maria caused significant rainfall, destruction, and mixing in the ocean waters. Hurricane Maria was a Category 4 hurricane when it transited directly over the island, passing through on 19–20 September. This hurricane caused the satellite SST data to switch from negative (cool) biases on the 19th (−0.60 °C = CoralTemp; −0.07 °C = G1SST; and −0.05 °C = OSTIA) to positive (warm) biases on the 20th (0.67 °C = CoralTemp; 1.20 °C = G1SST; and 0.78 °C = OSTIA; Figure 11).

According to the in situ data, the ocean temperature dropped ~1.5 °C from the 19th to the 20th, when the hurricane was passing. The warm bias continued until late February 2018, when ocean temperatures began to rise again, and the satellite SST biases returned to a negative (cool) state. The SST data warm biases seen after Hurricane Maria through late February could be attributed to increased water column mixing and seasonal cooler temperatures. Typically, La Parguera experiences low wave and tidal energy, and southeasterly winds from 3.1 to 7.7 m s^{-1} [39,48]. The reversal of the cool bias can be attributed to Hurricane Maria inducing higher seawater mixing for two weeks [49], and then the seasons transitioning to winter, causing temperatures to continue to decrease. From March 2018 to June 2019, SST data have a consistent negative (cool) bias. The lack of intense hurricanes in 2018 and the presence of a La Niña signal (e.g., MEI < −0.5 [50]) could be responsible for the more consistent pattern. NOAA reported lower than average rainfall for Puerto Rico over all of 2018. A weaker El Niño was also present near the end of 2018, continuing into spring 2019. NOAA reported less rainfall than usual for February 2019.

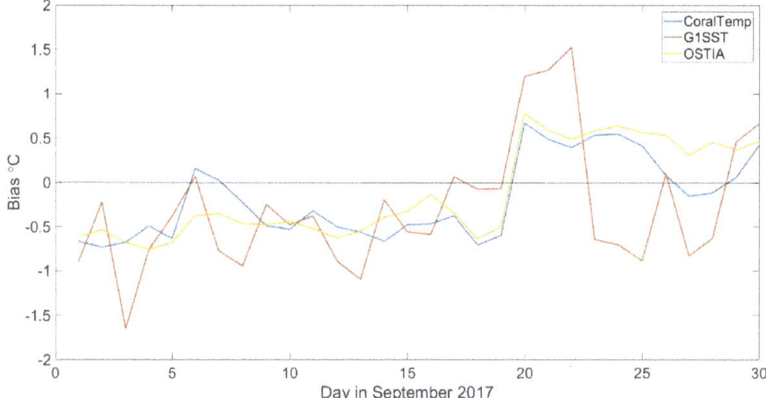

Figure 11. Time series of the bias (satellite—in situ) of the three remote sensing datasets for September 2017. All of the remote sensing datasets displayed a cooler bias, until Hurricane Maria transited the area on 19 September 2017, after which the biases became warm.

For San Cristobal, the JPL G1SST dataset yielded the smallest overall bias (−0.11 °C), but all three remote sensing products have high correlations with the in situ temperature measurements (CoralTemp = 0.95, OSTIA = 0.96, G1SST = 0.91). The same is true for Margarita Reef, where all the satellite SST datasets contained high correlations with the in situ data (CoralTemp = 0.95, OSTIA = 0.95, G1SST = 0.91). Observing the time series for all the study sites, there are two surprising temperature dips around December 2019, and January 2020, where the SST datasets are actually over predicting the temperature compared to the in situ loggers, and this was not seen in the previous years for the Cayo Enrique and Cayo Mario sites during these months. Generally, in winter, there was a closer match up observed between the SST data and in situ measurements. A possible explanation for these cool dips in the in situ temperature record that caused the satellites to have a warm bias, could be the magnitude 6.4 earthquake that occurred in southwestern Puerto Rico on 7 January 2020. This region hadn't experienced an earthquake of this magnitude since 1918 (USGS 2020). Scientists believe that the energy released during an earthquake could cause an increase in cold water anomalies, but further research is required to understand how local ocean temperatures change in response to earthquakes [51].

Inevitably, there will be some mismatch between the satellite SST and in situ measurements, because of the spatial scale, as the remote sensing modeled products combine data over a larger area (1–5 km pixel size), and the in situ data are point-based observations. Even though the satellite SST datasets underestimate the actual temperature at the depth of the coral for the majority of this study, the linear regression models with the satellite SST datasets as the predictor for the averaged in situ temperature logger measurements yielded an offset of ~1 °C. Cooler biases were also found in observations between satellite and in situ temperature measurements in previous studies [19,23], suggesting some consistency in bias across geographies. The occurrence of a cooler bias could be due to the ocean surface losing more heat at night to the atmosphere, which the temperature loggers at depth do not experience. Future studies could focus on heat budget analysis in corals in La Parguera, to assess if the coral tissues and reef framework contribute to the observed cool bias.

5. Conclusions

We sought to understand correspondence between temperatures observed at coral depth and those estimated for the sea surface by satellite-based remote sensing techniques, and to identify a statistical model capable of correcting bias in the remote sensing data, such that they may more accurately estimate temperatures at coral depths in the near shore zone. Since tropical corals live in coastal, subtidal areas that expose them to a wide range of temperature regimes, it is important to assess the satellite SST using in situ measurements to gain an accurate understanding of the temperature surrounding each coral reef ecosystem. It is also essential to explore the biases between satellite SST and in situ measurements, because NOAA's CRW uses satellite SST to produce their products, which are used by coral managers to inform them when their reefs might experience bleaching. Overall, a strong positive correlation was observed between all the satellite products and in situ measurements, with no real differences found between logger sites and depth, and this would be expected in such shallow, well-mixed waters. However, a consistent negative (cool) bias was found between the in situ temperature data and satellite SST datasets during the warmer months with a closer match between them during the colder months. The warm season biases for the satellite SST datasets were all around 1 °C, providing a good overall agreement between the satellite SST and the in situ loggers. While other studies have suggested similar biases in different locations, the spatial coherence of systematic biases between surface and reef-depth temperatures has not been fully explored. Expanding this study with a broader network of in situ sensors to simultaneously evaluate more SST pixels is needed, especially around Puerto Rico, and a longer time series is required.

Author Contributions: A.M.G. developed the project plan with support from K.C.M. and K.S. in an advisory role. A.M.G. analyzed the satellite data and wrote the manuscript. A.M.G., M.C., K.C.M., and S.D. deployed the in situ sensors, and provided supporting SCUBA fieldwork. K.C.M., K.S. and S.D. provided manuscript edits. R.A.A. and W.J.H. helped with the data collection and manuscript edits. All authors have read and agreed to the published version of the manuscript.

Funding: This study is supported and monitored by The National Oceanic and Atmospheric Administration –Cooperative Science Center for Earth System Sciences and Remote Sensing Technologies (NOAA-CESSRST) under the Cooperative Agreement Grant # NA16SEC4810008. The authors would like to thank The City College of New York, NOAA-CESSRST (aka CREST) program and NOAA Office of Education, Educational Partnership Program for full fellowship support for Andrea Gomez. The statements contained within the manuscript/research article are not the opinions of the funding agency or the U.S. government, but reflect the author's opinions. This research is also partly funded by The Doctoral Research Student Grant (DRSG) and the Provost's Pre-dissertation Summer Research Award, both of which come from The CUNY Graduate Center.

Acknowledgments: We thank Milton Carlo for his tremendous help with the logger deployment and data collection, and the University of Puerto Rico, Mayagüez Bio-Optical Oceanography, Remote Sensing lab for use of their facilities. We also thank Mark Eakin, and Jacqueline De La Cour from NOAA's Coral Reef Watch for their help editing the manuscript. Finally we thank Aaron Davitt, Brian Lamb, Derek Tesser, Kat Jensen, Michael Brown, and Jessica Rosenqvist from the Ecosystem Science Lab for all their tremendous support and feedback, and Adrian Diaz and Sonia Dagan for their help with Matlab and the statistical analyses. The HOBO in situ temperature data supporting the conclusions can be accessed at the TemperateReefBase Data Portal (https://temperatereefbase.imas.utas.edu.au/static/landing.html).

References

1. Hoegh-Guldberg, O.; Mumby, P.J.; Hooten, A.J.; Steneck, R.S.; Greenfield, P.; Gomez, E.; Harvell, C.D.; Sale, P.F.; Edwards, A.J.; Caldeira, K.; et al. Coral Reefs Under Rapid Climate Change and Ocean Acidification. *Science* **2007**, *318*, 1737–1742. [CrossRef] [PubMed]

2. Anthony, K.R.N. Coral Reefs Under Climate Change and Ocean Acidification: Challenges and Opportunities for Management and Policy. *Annu. Rev. Environ. Resour.* **2016**, *41*, 59–81. [CrossRef]

3. Van Hooidonk, R.; Maynard, J.; Tamelander, J.; Gove, J.; Ahmadia, G.; Raymundo, L.; Williams, G.; Heron, S.F.; Planes, S. Local-scale projections of coral reef futures and implications of the Paris Agreement. *Sci. Rep.* **2016**, *6*, 39666. [CrossRef]

4. Coles, S.; Brown, B.E. Coral bleaching—capacity for acclimatization and adaptation. *Adv. Mar. Biol.* **2003**, *46*, 183–223. [CrossRef] [PubMed]

5. Brown, B.E.; Dunne, R.P. *Coral Bleaching: The Roles of Sea Temperature and Solar Radiation*; Woodley, C.M., Downs, C.A., Bruckner, A.W., Porter, J.W., Galloway, S.B., Eds.; John Wiley & Sons, Inc.: Hoboken, NJ, USA, 2016; pp. 266–283.

6. Downs, C.A.; Fauth, J.E.; Halas, J.C.; Dustan, P.; Bemiss, J.; Woodley, C.M. Oxidative stress and seasonal coral bleaching. *Free. Radic. Boil. Med.* **2002**, *33*, 533–543. [CrossRef]

7. Cunning, R.; Muller, E.B.; Gates, R.D.; Nisbet, R.M. A dynamic bioenergetic model for coral- Symbiodinium symbioses and coral bleaching as an alternate stable state. *J. Theor. Boil.* **2017**, *431*, 49–62. [CrossRef]

8. Hughes, T.P.; Kerry, J.T.; Baird, A.H.; Connolly, S.R.; Dietzel, A.; Eakin, C.M.; Heron, S.F.; Hoey, A.S.; Hoogenboom, M.O.; Liu, G.; et al. Global warming transforms coral reef assemblages. *Nature* **2018**, *556*, 492–496. [CrossRef]

9. Eakin, C.; Liu, G.; Gomez, A.; De La Cour, J.; Heron, S.; Skirving, W. Ding, Dong, the Witch is Dead (?)-Three Years of Global Coral Bleaching 2014–2017. *Reef Encount.* **2017**, *45*, 33–38.

10. Camp, J.; Scaife, A.A.; Heming, J. Predictability of the 2017 North Atlantic hurricane season. *Atmos. Sci. Lett.* **2018**, *19*, e813. [CrossRef]

11. Reardon, S. Hurricane Maria's wrath leaves clues to coral reefs' future. *Nature* **2018**, *560*, 421–422. [CrossRef]

12. Rowlands, G.; Purkis, S.; Riegl, B.; Metsamaa, L.; Bruckner, A.; Renaud, P. Satellite imaging coral reef resilience at regional scale. A case-study from Saudi Arabia. *Mar. Pollut. Bull.* **2012**, *64*, 1222–1237. [CrossRef] [PubMed]

13. Roelfsema, C.; Phinn, S. Integrating field data with high spatial resolution multispectral satellite imagery for calibration and validation of coral reef benthic community maps. *J. Appl. Remote. Sens.* **2010**, *4*, 043527. [CrossRef]

14. Hedley, J.D.; Roelfsema, C.; Chollett, I.; Harborne, A.; Heron, S.F.; Weeks, S.; Skirving, W.J.; Strong, A.E.; Eakin, C.M.; Christensen, T.R.L.; et al. Remote Sensing of Coral Reefs for Monitoring and Management: A Review. *Remote. Sens.* **2016**, *8*, 118. [CrossRef]

15. Liu, G.; Skirving, W.J.; Geiger, E.F.; De La Cour, J.L.; Marsh, B.L.; Heron, S.F.; Tirak, K.V.; Strong, A.E.; Eakin, C.M. NOAA Coral Reef Watch's 5km Satellite Coral Bleaching Heat Stress Monitoring Product Suite Version 3 and Four-Month Outlook Version 4. *Reef Encount.* **2017**, *45*, 39–45.

16. Strong, A.E.; Liu, G.; Skirving, W.; Eakin, C.M. NOAA's Coral Reef Watch program from satellite observations. *Ann. GIS* **2011**, *17*, 83–92. [CrossRef]

17. Montgomery, R.S.; Strong, A.E. Coral bleaching threatens oceans, life. *Eos* **1994**, *75*, 145. [CrossRef]

18. McClanahan, T.; Ateweberhan, M.; Sebastián, C.R.; Graham, N.A.J.; Wilson, S.K.; Bruggemann, J.H.; Guillaume, M. Predictability of coral bleaching from synoptic satellite and in situ temperature observations. *Coral Reefs* **2007**, *26*, 695–701. [CrossRef]

19. Castillo, K.D.; Lima, F.P. Comparison of in situ and satellite-derived (MODIS-Aqua/Terra) methods for assessing temperatures on coral reefs. *Limnol. Oceanogr. Methods* **2010**, *8*, 107–117. [CrossRef]

20. Stobart, B.; Mayfield, S.; Mundy, C.; Hobday, A.J.; Hartog, J.R. Comparison of in situ and satellite sea surface-temperature data from South Australia and Tasmania: how reliable are satellite data as a proxy for coastal temperatures in temperate southern Australia? *Mar. Freshw. Res.* **2016**, *67*, 612. [CrossRef]

21. Lathlean, J.; Ayre, D.; Minchinton, T. Rocky intertidal temperature variability along the southeast coast of Australia: comparing data from in situ loggers, satellite-derived SST and terrestrial weather stations. *Mar. Ecol. Prog. Ser.* **2011**, *439*, 83–95. [CrossRef]

22. Wu, Y.; Sheng, J.; Senciall, D.; Tang, C. A comparative study of satellite-based operational analyses and ship-based in-situ observations of sea surface temperatures over the eastern Canadian shelf. *Satell. Oceanogr. Meteorol.* **2016**, *1*, 29–38. [CrossRef]

23. Thakur, K.; Vanderstichel, R.; Barrell, J.; Stryhn, H.; Patanasatienkul, T.; Revie, C.W. Comparison of Remotely-Sensed Sea Surface Temperature and Salinity Products With in Situ Measurements From British Columbia, Canada. *Front. Mar. Sci.* **2018**, *5*, 121. [CrossRef]

24. Smit, A.J.; Roberts, M.; Anderson, R.J.; Dufois, F.; Dudley, S.F.J.; Bornman, T.; Olbers, J.; Bolton, J.J. A Coastal Seawater Temperature Dataset for Biogeographical Studies: Large Biases between In Situ and Remotely-Sensed Data Sets around the Coast of South Africa. *PLoS ONE* **2013**, *8*, e81944. [CrossRef] [PubMed]

25. Brewin, R.J.W.; Smale, D.A.; Moore, P.J.; Nencioli, F.; Miller, P.; Taylor, B.H.; Smyth, T.; Fishwick, J.R.; Yang, M.X. Evaluating Operational AVHRR Sea Surface Temperature Data at the Coastline Using Benthic Temperature Loggers. *Remote. Sens.* **2018**, *10*, 925. [CrossRef]

26. Donner, S.D.; Skirving, W.; Little, C.M.; Oppenheimer, M.; Hoegh-Guldberg, O. Global assessment of coral bleaching and required rates of adaptation under climate change. *Glob. Chang. Boil.* **2005**, *11*, 2251–2265. [CrossRef]

27. Strong, A.E.; Arzayus, F.; Skirving, W.; Heron, S.F. Identifying coral bleaching remotely via coral reef watch—Improved integration and implications for changing climate. *Coast. Estuar. Stud.* **2006**, *61*, 163–180. [CrossRef]

28. Chao, Y.; Li, Z.; Farrara, J.D.; Hung, P. Blending Sea Surface Temperatures from Multiple Satellites and In Situ Observations for Coastal Oceans. *J. Atmospheric Ocean. Technol.* **2009**, *26*, 1415–1426. [CrossRef]

29. Donlon, C.; Martin, M.J.; Stark, J.; Roberts-Jones, J.; Fiedler, E.; Wimmer, W. The Operational Sea Surface Temperature and Sea Ice Analysis (OSTIA) system. *Remote. Sens. Environ.* **2012**, *116*, 140–158. [CrossRef]

30. Winter, A.; Appeldoorn, R.S.; Bruckner, A.W.; Williams, E.H., Jr. Sea surface temperatures and coral reef bleaching off La Parguera, Puerto Rico (northeastern Caribbean Sea). *Coral Reefs* **1998**, *17*, 377–382. [CrossRef]

31. Ryan-Mishkin, K.; Walsh, J.P.; Corbett, D.R.; Dail, M.B. Modern Sedimentation in a Mixed Siliciclastic-Carbonate Coral Reel Environment, La Parguera, Puerto Rico. *Caribb. J. Sci.* **2009**, *45*, 151–167. [CrossRef]

32. Appeldoorn, R.S.; Ballantine, D.; Bejarano, I.; Carlo, M.; Nemeth, M.; Otero, E.; Pagan, F.; Ruiz, H.; Schizas, N.; Sherman, C.; et al. Mesophotic coral ecosystems under anthropogenic stress: A case study at Ponce, Puerto Rico. *Coral Reefs* **2015**, *35*, 63–75. [CrossRef]

33. Kilbourne, K.H.; Quinn, T.M.; Webb, R.; Guilderson, T.; Nyberg, J.; Winter, A. Paleoclimate proxy perspective on Caribbean climate since the year 1751: Evidence of cooler temperatures and multidecadal variability. *Paleoceanogr. Paleoclimatol.* **2008**, *23*, 23. [CrossRef]

34. Winter, A.; Paul, A.; Nyberg, J.; Oba, T.; Lundberg, J.; Schrag, D.; Taggart, B. Orbital control of low-latitude seasonality during the Eemian. *Geophys. Res. Lett.* **2003**, *30*, 30. [CrossRef]

35. Hu, A.; Meehl, G.A.; Han, W. Detecting thermohaline circulation changes from ocean properties in a coupled model. *Geophys. Res. Lett.* **2004**, *31*. [CrossRef]

36. Eakin, C.M.; Sweatman, H.P.A.; Brainard, R.E. The 2014–2017 global-scale coral bleaching event: insights and impacts. *Coral Reefs* **2019**, *38*, 539–545. [CrossRef]

37. Hallam, S.; Marsh, R.; Josey, S.A.; Hyder, P.; Moat, B.; Hirschi, J.J.-M. Ocean precursors to the extreme Atlantic 2017 hurricane season. *Nat. Commun.* **2019**, *10*, 896. [CrossRef]

38. Ballantine, D.L.; Appeldoorn, R.S.; Yoshioka, P.; Weil, E.; Armstrong, R.; Garcia, J.R.; Otero, E.; Pagan, F.; Sherman, C.; Hernandez-Delgado, E.A.; et al. Biology and Ecology of Puerto Rican Coral Reefs. In *Coral Reefs of the USA*; Springer Science and Business Media LLC: Berlin/Heidelberg, Germany, 2008; Volume 1, pp. 375–406.

39. Fitt, W.K.; McFarland, F.K.; Warner, M.E.; Chilcoat, G.C. Seasonal patterns of tissue biomass and densities of symbiotic dinoflagellates in reef corals and relation to coral bleaching. *Limnol. Oceanogr.* **2000**, *45*, 677–685. [CrossRef]

40. Rodolfo-Metalpa, R.; Richard, C.; Allemand, D.; Bianchi, C.N.; Morri, C.; Ferrier-Pagès, C. Response of zooxanthellae in symbiosis with the Mediterranean corals Cladocora caespitosa and Oculina patagonica to elevated temperatures. *Mar. Boil.* **2006**, *150*, 45–55. [CrossRef]

41. Oliver, T.A.; Palumbi, S.R. Many corals host thermally resistant symbionts in high-temperature habitat. *Coral Reefs* **2010**, *30*, 241–250. [CrossRef]

42. DeBose, J.L.; Nuttall, M.F.; Hickerson, E.L.; Schmahl, G.P. A high-latitude coral community with an uncertain future: Stetson Bank, northwestern Gulf of Mexico. *Coral Reefs* **2012**, *32*, 255–267. [CrossRef]

43. Camp, E.F.; Smith, D.J.; Evenhuis, C.; Enochs, I.; Manzello, D.; Woodcock, S.; Suggett, D.J. Acclimatization to high-variance habitats does not enhance physiological tolerance of two key Caribbean corals to future temperature and pH. In Proceedings of the Royal Society B: Biological Sciences; The Royal Society: London, UK, 2016; Volume 283, p. 20160442.

44. Johnston, M.A.; Hickerson, E.L.; Nuttall, M.F.; Blakeway, R.D.; Sterne, T.K.; Eckert, R.J.; Schmahl, G.P. Coral bleaching and recovery from 2016 to 2017 at East and West Flower Garden Banks, Gulf of Mexico. *Coral Reefs* **2019**, *38*, 787–799. [CrossRef]

45. Sreejith, O.P.; Shenoi, S.S.C. An evaluation of satellite and in situ based sea surface temperature datasets in the North Indian Ocean region. *Int. J. Remote. Sens.* **2002**, *23*, 5165–5175. [CrossRef]

46. Hauke, J.; Kossowski, T. Comparison of values of Pearson's and Spearman's correlation coefficients on the same sets of data. *Quaest. Geogr.* **2011**, *30*, 87–93. [CrossRef]

47. Sarmiento, J.L.; Slater, R.; Barber, R.; Bopp, L.; Doney, S.C.; Hirst, A.C.; Kleypas, J.; Matear, R.J.; Mikolajewicz, U.; Monfray, P.; et al. Response of ocean ecosystems to climate warming. *Glob. Biogeochem. Cycles* **2004**, *18*, 18. [CrossRef]

48. Warne, A.G.; Webb, R.M.; Larsen, M.C. *Water, Sediment, and Nutrient Discharge Characteristics of Rivers in Puerto Rico, and Their Potential Influence on Coral Reefs*; US Department of the Interior, US Geological Survey: Reston, VA, USA, 2005.

49. Hu, A.; Meehl, G.A. Effect of the Atlantic hurricanes on the oceanic meridional overturning circulation and heat transport. *Geophys. Res. Lett.* **2009**, *36*, 36. [CrossRef]

50. Wolter, K.; Timlin, M.S. Measuring the strength of ENSO events: How does 1997/98 rank? *Weather* **1998**, *53*, 315–324. [CrossRef]

51. Nosov, M.A. Ocean surface temperature anomalies from underwater earthquakes. *Oceanogr. Lit. Rev.* **1998**, *9*, 1491.

Journal of
Marine Science and Engineering

Article

Increasing Trends in Air and Sea Surface Temperature in the Central Adriatic Sea (Croatia)

Ognjen Bonacci [1], Duje Bonacci [2], Matko Patekar [3,*] and Marco Pola [3]

[1]   Faculty of Civil Engineering, Architecture and Geodesy, University of Split, Matice Hrvatske 15, 21000 Split, Croatia; obonacci@gradst.hr

[2]   Faculty of Croatian Studies, University of Zagreb, Borongajska Cesta 83d, 10000 Zagreb, Croatia; dbonacci@voxpopuli.hr

[3]   Department of Hydrogeology and Engineering Geology, Croatian Geological Survey, Sachsova 2, 10000 Zagreb, Croatia; mpola@hgi-cgs.hr

*   Correspondence: mpatekar@hgi-cgs.hr; Tel.: + 385-98-904-26-99

Abstract: The Adriatic Sea and its coastal region have experienced significant environmental changes in recent decades, aggravated by climate change. The most prominent effects of climate change (namely, an increase in sea surface and air temperature together with changes in the precipitation regime) could have an adverse effect on social and environmental processes. In this study, we analyzed the time series of sea surface temperature and air temperature measured at three meteorological stations in the Croatian part of the Adriatic Sea. To assess the trends and variations in the time series of sea surface and air temperature, different statistical methods were employed, i.e., linear and quadratic regressions, Mann–Kendall test, Rescaled Adjusted Partial Sums method, and autocorrelation. The results evidenced increasing trends in the mean annual sea surface temperature and air temperature; furthermore, sudden variations in values were observed in 1998 and 1992, respectively. Increasing trends in the mean monthly sea surface temperature and air temperature occurred in the warmer parts of the year (from March to August). The results of this study could provide a foundation for stakeholders, decision–makers, and other scientists for developing effective measures to mitigate the negative effects of climate change in the scattered environment of the Adriatic islands and coastal region.

Keywords: sea surface temperature; air temperature; Mann–Kendall test; Split; Hvar; Komiža

Citation: Bonacci, O.; Bonacci, D.; Patekar, M.; Pola, M. Increasing Trends in Air and Sea Surface Temperature in the Central Adriatic Sea (Croatia). *J. Mar. Sci. Eng.* **2021**, 9, 358. https://doi.org/10.3390/jmse9040358

Academic Editor: Francisco Pastor Guzman

Received: 30 January 2021
Accepted: 23 March 2021
Published: 26 March 2021

Publisher's Note: MDPI stays neutral with regard to jurisdictional claims in published maps and institutional affiliations.

1. Introduction

The Adriatic Sea is the northernmost part of the Mediterranean Sea and represents a vast bay that is indented far into the European mainland. The surface of the Adriatic Sea is 138,595 km^2, which constitutes 4.6% of the total surface of the Mediterranean Sea. The Adriatic Sea stretches in NW–SE direction for 870 km, with an average width of approximately 200 km. It is connected to the Mediterranean Sea by the 70 km–wide Strait of Otranto. The Adriatic Sea is heterogeneous in terms of physical, chemical, and biological properties [1]. Based on the average depth, the Adriatic is divided into (i) very shallow Northern Adriatic (<35 m), (ii) Central Adriatic Sea (~140 m), and (iii) Southern Adriatic with an average depth of >200 m and maximum depth up to 1228 m [2]. The majority of the islands in the Adriatic Sea are within the Croatian territory. The total length of the Croatian coast is 5835 km, i.e., the lengths of the island and mainland coast are 4058 km and 1777 km, respectively.

One of the key environmental questions is how climate change, particularly global warming, will affect the environmental and social processes in different regions of the planet. The investigation of the effects of climate change in the coastal environment is a particularly complex task because of the simultaneous influence of vast land and water masses, which react differently to climate change. Since climate change continues to

49

intensify, both the Adriatic and the Mediterranean regions are considered hot spots of climate change. In particular, the Adriatic ecosystems suffer the combined effect of climate change in the Northern hemisphere and local climate variations [3]. Regional atmospheric oscillations, as a result of the air pressure variations in the Northern hemisphere, and the intensification of water inflow from the Eastern Mediterranean affected the Adriatic Sea temperature and salinity. Further local factors impacting the Adriatic marine ecosystems are: (i) increasing sea surface temperature, (ii) negative trends in the precipitation patterns, particularly in the winter season, and (iii) reduced inflow of fresh water and nutrients as a result of the decreasing inflow from the Po river [4]. This complex situation is fostered by the variable geometry and topography of the Adriatic region where each bay, island, or channel in the Adriatic Sea is site–specific in terms of oceanographic properties [5] and reactions to climate change [6,7]. The devastation of the environment, the increasing demand for fresh water for potable and irrigation purposes, and the changes in the already unfavorable water balances of the Mediterranean semi–arid climate could lead to adverse environmental and social consequences in the near future. Commonly, the karst environments are scarce in surface hydrography and groundwater is the main freshwater resource. Hence, the Mediterranean water resources are under significant stress due to over–abstraction, climate change, and the high possibility of seawater intrusions into karst aquifers in the coastal areas or the islands.

To understand the trends and the behavior of the air temperature on small islands and in the coastal areas, it is necessary to understand the interaction between the sea surface temperature and the air temperature. Vlahakis and Pollatou [8] emphasized that the sea surface temperature is the key factor in the assessment of the climate and climate–related processes on all scales, especially on the islands and in the coastal areas. Sea surface temperature has a direct influence on global energy transfer, atmospheric processes, precipitation, evapotranspiration, moisture in air and soil, wind development, hydrological cycle, and other ecological or social processes [9]. However, sea surface temperature is not considered a standard meteorological parameter even though its effect on the atmosphere is immense. Furthermore, the time series of the sea surface temperature are considerably shorter when compared to the air temperature. The sea has a higher thermal capacity than the land, resulting in slower heating and cooling. During winter and summer, the sea acts as a buffer that moderates the air temperature, resulting in a lower range of the sea surface temperature than the range of air temperature over land [10]. The flow of the seawater and other turbulent processes cause the deeper layers of the sea to mix with the surficial ones, influencing the air temperature. The air temperature reacts differently to changes in the sea surface temperature, and this is particularly pronounced on the small islands and in the coastal areas. Therefore, it is necessary to analyze each location based on a detailed and reliable time series of measured data.

Time series of the sea surface temperature and the air temperature measured at three meteorological stations located on Central Adriatic islands (Hvar and Vis) and on the mainland (Split) were analyzed in this study. Firstly, we will describe the study area and the available dataset. The dataset will be analyzed using different statistical approaches and the obtained results will be used to assess the occurrence of trends within the time series of the sea surface temperature and the air temperature. These data will provide a foundation for stakeholders, decision–makers, and other scientists for developing effective measures to mitigate the negative effects of climate change in the Adriatic region and beyond.

2. Study Area

The study area is located in the central part of the Adriatic Sea and consists of the two meteorological stations on the islands of Hvar and Vis, and the meteorological station in the city of Split on the mainland. The map of the study area showing the analyzed meteorological stations is in Figure 1.

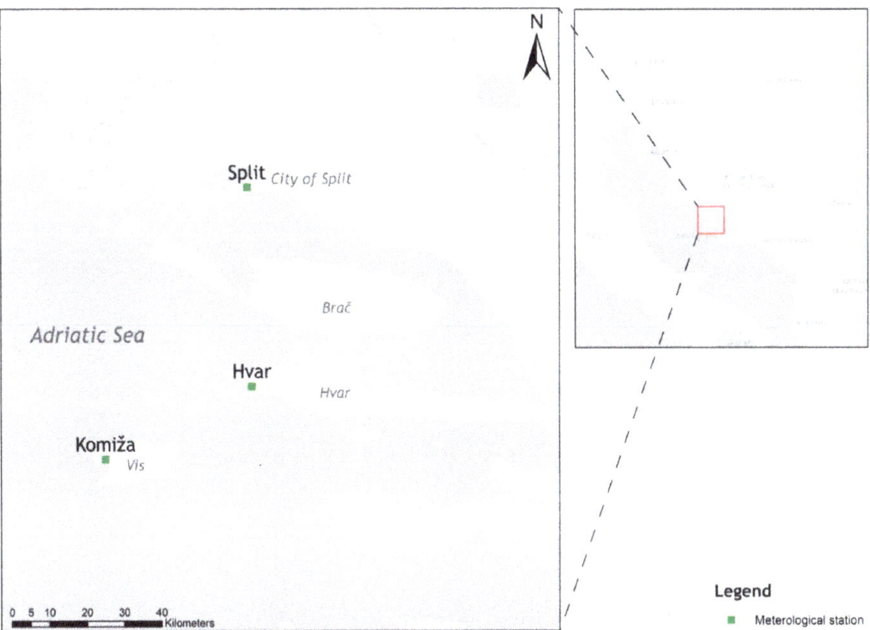

Figure 1. Map of the study area showing the locations of the three meteorological stations whose data were analyzed in this study.

The main characteristics of the analyzed stations used in this study are summarized in Table 1.

Table 1. Characteristics of the investigated meteorological stations.

	Split	Hvar	Komiža
Latitude	43°30′30″	43°10′16″	43°02′54″
Longitude	16°25′35″	16°26′13″	16°05′07″
H (m a.s.l.)	122	20	20
Investigated period (SST)	1960–2019	1964–2019	1991–2019
Investigated period (T_A)	1960–2019	1960–2019	1960–2019
L (m)	3100	100	190
L_s (m)	515	80	175

Note: H—the altitude of the meteorological station; SST—sea surface temperature; T_A—air temperature; L—the distance between the SST and the air temperature measuring point; L_s—minimum distance between meteorological stations and the seashore.

The meteorological station Split is located in the eastern part of Split, on the Marjan peninsula. Marjan is a forest park used for recreational purposes. The meteorological station is in the vicinity of the second tallest peak on Marjan (178 m a.s.l.). In 2011, the population of Split was 178,192. Despite rapid urbanization of the Split agglomeration, the meteorological station Split is not affected by urbanization due to its position within a protected forest area. The sea surface temperature (SST hereafter) is measured at the endpoint of the pier at the easternmost point of the Marjan peninsula.

The small island of Hvar belongs to the Middle–Dalmatian island group. With a surface area of 297.32 km^2 and a coastline of 270 km, Hvar is the fourth biggest island in the Croatian part of the Adriatic Sea [11]. In 2011, the city of Hvar had a population of 4251. The meteorological station Hvar is situated in a small forest grove, away from the urbanized city center. The SST is measured at the endpoint of a small pier 100 m from the meteorological station.

Vis is a small remote island in the Adriatic Sea. Its distance from the mainland is 43 km and the island is exposed to very strong winds. With a surface area of 89.72 km^2 and coastline of 85 km, Vis is the ninth biggest island in the Croatian part of the Adriatic Sea [11]. The meteorological station is located in the city of Komiža, on the northern edge of the urbanized area. In 2011, Komiža had a population of 1526. The SST is measured at the endpoint of a pier in Komiža, 190 m from the meteorological station. Vis Island has favorable geological and hydrogeological conditions that have enabled the formation of high–quality karst aquifers from which the fresh water is abstracted. Hence, Vis Island and its groundwater resources are considered highly vulnerable to climate change [7].

According to Köppen–Geiger climate classification, the study area is characterized by the Csa climate type [12], sometimes called the "olive" climate. It is a semiarid variety of Mediterranean climate characterized by mild and humid winters, and dry and hot summers.

3. Materials and Methods

3.1. Data Collection

The SST and the air temperature data used in this study were provided by the Croatian Meteorological and Hydrological Service (DHMZ). The conventional method of measuring the SST is performed using a thermometer immersed in the sea at a depth of 30 cm, given that the sea is not shallower than 180 cm at the measuring point. The thermometer is immersed for three minutes, and after that, it is taken out and read quickly [13]. The SST is measured three times a day at 7, 14, and 21 h local time. The measurement of the SST is performed by personnel from the nearby meteorological station. Values of the mean monthly and the mean annual SST were analyzed in this study.

Furthermore, the time series of air temperature were measured at main meteorological stations. An interesting fact is that the meteorological station Hvar started to operate in 1858, followed by the Split station a year after, and Komiža in 1959. The World Meteorological Organization (WMO) recognized the quality of the station in Hvar and awarded it a Centennial Observing Station title. The air temperature is measured hourly at a standard height of 2 m above the ground. In this study, the mean monthly and the mean annual air temperature for the period 1960–2019 were analyzed. Despite several approaches (e.g., [14–16]), the method employed by the DHMZ [13], which is the most common in Europe, defines the mean daily air temperature as:

$$t_{mean,daily} = \frac{t_7 + t_{14} + 2t_{21}}{4},$$ (1)

where t_7, t_{14}, and t_{21} are the air temperature values measured at 7, 14, and 21 h (local time), respectively. The same procedure is applied for the SST.

3.2. Statistical Methods

Linear and quadratic regressions were performed on the time series of mean monthly and mean annual SST and air temperature from three stations analyzed in this study. The linear regression equation is given as:

$$T = (a \times t) + b,$$ (2)

and the quadratic regression equation as:

$$T = (c \times t^2) + (d \times t) + e,$$ (3)

where T is the mean monthly or mean annual SST or air temperature in year t, a and b are linear regression coefficients, and c, d, and e are quadratic regression coefficients. All five coefficients are calculated by the least–squares method. The coefficient a represents the slope of the regression line whose dimension is °C/year, and it is the indicator of the average intensity of the increasing or decreasing trends in the values of the analyzed time series. The correlation coefficients r^2 and R^2 were calculated for the linear and the quadratic regressions, respectively. Both coefficients show the strength and the direction of linear and quadratic correlation between variables x (time) and y (the mean monthly or the mean annual SST or air temperature).

To assess whether the time series have monotonic increasing or decreasing trends, the Mann–Kendall (M–K hereafter) non–parametric test was used [17]. The null hypothesis for this test is that there is no monotonic trend within the analyzed time series, while the alternate hypothesis is that the trend exists. As a criterion to accept the alternate hypothesis (i.e., the presence of an increasing or decreasing trend), p–values < 0.05 were used in this study.

Furthermore, the Rescaled Adjusted Partial Sums method (RAPS hereafter) was used to detect statistically significant peaks or declines in values (i.e., the trends variations) within the analyzed time series [18,19]. This method allows overcoming random and irregular fluctuations as well as rough errors in values within the time series, which may be hidden from the common plots of values of the time series. Based on the RAPS results, sub–periods with similar characteristics or a larger number of trends within the time series could be distinguished. The formula for the calculation of RAPS is:

$$RAPS_k = \sum_{t=1}^{k} \frac{Y_t - Y_m}{S_y},\qquad(4)$$

where Y_t is the value of the observed parameter at time t, Y_m is the mean value of observed time series, S_y is the standard deviation of the observed time series, and k is the number of observations. The breakpoints between the sub–periods were established when the trend of RAPS results showed a significant variation.

The differences in statistical parameters of the two neighboring sub–periods defined by the RAPS were evaluated by the F–test and the t–test [20]. In particular, the F–test was used to assess the equality of variances between the two normally distributed populations (i.e., sub–periods). The t–test was used to determine whether there is a statistical difference between the mean values of the two sub–periods. In this study, both tests accept the null hypothesis for p–values < 0.05.

Furthermore, the autocorrelation of the time series was determined. Autocorrelation is a mathematical function representing the degree of similarity between the specific time series and a lagged version of the same time series over successive time intervals. Auto-correlation coefficient r ranges from -1 to 1 and it measures the strength of a relationship between the current value of the variable with its shifted value. In this study, the interval of the shifting variable was set to 1 year. For $r < 0.2$, the time series is not autocorrelated meaning that the behavior of the values does not depend on the previous values [21].

4. Results and Discussion

In this chapter, temporal changes in the mean annual and the mean monthly SST and air temperature were analyzed. The analyzed time series are not identical in duration, which will affect the reliable comparison of the results obtained at three analyzed stations to a minor extent. Despite the differences in the duration of the analyzed time series, contemporaneous time series were available for the last 29–year period (i.e., 1991–2019), when the most significant increases in the SST and the air temperature were observed. This fact will allow a more reliable conclusion about the recent and future behavior of the SST and the air temperature in the study area.

4.1. Analyses of the Mean Annual Sea Surface Temperature and Air Temperature

Table 2 shows a summary of the exploratory statistical analysis (minimum, average, maximum, and range) of the mean annual SST, the air temperature (T_A), and their difference ($\Delta T = SST-T_A$) at the analyzed stations. The longest time series analyzed in this study were recorded in Split (from 1960 to 2019), followed by Hvar (from 1964 to 2019), and Komiža with the shortest time series (from 1991 to 2019).

Table 2. Statistics (minimum, average, maximum) of the mean annual SST, air temperature (T_A), and their difference ($\Delta T = SST-T_A$).

T (°C)		SPLIT	HVAR	KOMIŽA
		1960−2019	1964−2019	1991−2019
SST	Minimum	16.3	17.0	17.9
	Average	17.4	18.1	18.9
	Maximum	18.7	19.3	19.5
	Range	2.4	2.3	1.6
T_A	Minimum	15.1	15.5	16.2
	Average	16.4	16.7	17.2
	Maximum	17.8	18.2	18.0
	Range	2.7	2.7	1.8
ΔT	Minimum	1.2	1.5	1.7
	Average	1.1	1.4	1.7
	Maximum	0.9	1.1	1.5
	Range	-0.3	-0.4	-0.2

The minimum, average, and maximum values of the mean annual SST measured in Split were 16.3 °C, 17.4 °C, and 18.7 °C, respectively. The values measured at Hvar were slightly higher being 17 °C, 18.1 °C, and 19.3 °C, respectively. Komiža had the highest values of SST, with the minimum, average, and maximum mean annual SST being 17.9 °C, 18.9 °C, and 19.5 °C, respectively. The range of the mean annual SST was 2.4 °C, 2.3 °C, and 1.6 °C in Split, Hvar, and Komiža, respectively. The lowest range of the mean annual SST in Komiža reflects its furthest position in the open sea among the analyzed stations.

The distribution of the air temperature (T_A) values was similar to the SST in terms of Split having the lowest values and Komiža the highest. The minimum, average, and maximum values of the mean annual air temperature measured in Split were 15.1 °C, 16.4 °C, and 17.8 °C, respectively; slightly higher values were measured in Hvar as 15.5 °C, 16.7 °C, and 18.2 °C, respectively. In Komiža, the minimum, average, and maximum values of the mean annual air temperature were 16.2 °C, 17.2 °C, and 18 °C. Hvar and Split had an identical range of the mean annual air temperature, 2.7 °C, while the range in Komiža was 1.8 °C.

The ΔT values (SST-T_A) showed the same distribution as the SST and the T_A. The ΔT of the minimum values were 1.2 °C, 1.5 °C, and 1.7 °C, in Split, Hvar, and Komiža, respectively. The ΔT average values were 1.1 °C, 1.4 °C, and 1.7 °C, while the ΔT maximum values were 0.9 °C, 1.1 °C, and 1.5 °C at the same stations, respectively. The positive values of ΔT indicated that the mean annual SST was always higher than the mean annual air temperature. It should be noted that ΔT of the minimum values were higher than the ΔT of the maximum values due to the smaller amplitude of the SST than the air temperature.

Figure 2 shows the time series of the mean annual SST measured at Split (shown in blue), Hvar (shown in green), and Komiža (shown in red) stations.

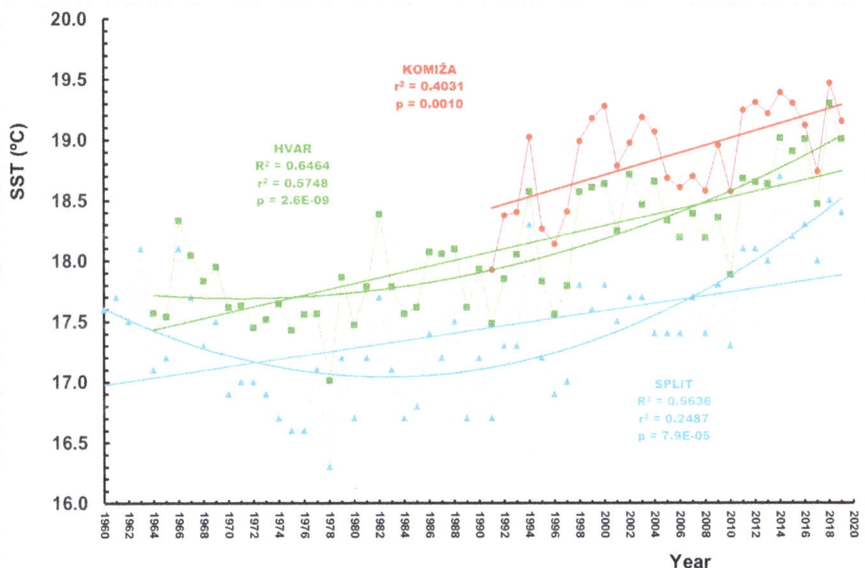

Figure 2. Time series of the mean annual SST measured at Split, Hvar, and Komiža stations. The r^2 and R^2 represent the r–squared values of the correlation coefficients of the linear and quadratic regressions, respectively, and p represents the Mann–Kendall (M–K) test values.

The linear regressions evidenced increasing trends in the SST at all analyzed stations, which were corroborated by the results of the M–K test (i.e., $p < 0.01$). However, to achieve better fitting to the data, quadratic regressions were performed on the time series of Split and Hvar stations. The results showed slightly higher R^2 values than the coefficient of linear regressions r^2. Quadratic regression was not performed on the time series from Komiža due to the missing data before 1991.

The RAPS method has been used on the time series of SST from all analyzed stations. The results are in the Supplementary Material (S1) and they evidenced the presence of two sub–periods: (i) from the beginning of the respective time series until 1997, and (ii) from 1998 to 2019 (Figure 3).

Despite the differences in the duration of the time series, increasing (Hvar and Komiža) or decreasing (Split) trends in SST were not statistically significant in the first sub–period defined by the RAPS (p–values > 0.05; Figure 3). Statistically significant increasing trends in the SST occurred in the second sub–period at stations in Hvar and Split (p–values < 0.05), and in Komiža there was not a statistically significant trend in the second sub–period (p–values > 0.05).

The average values of the mean annual SST within the sub–periods defined by the RAPS method are shown in Figure 3 and Table 3. The statistical analyses evidenced statistically significant differences between the average values of the mean annual SST in the two sub–periods, with the rejection of the null hypothesis of the t–test (low p–values, i.e., $p < 0.01$), and similar variances of the sub–periods reflecting the failure to reject the null hypothesis of the F–test (high p–value).

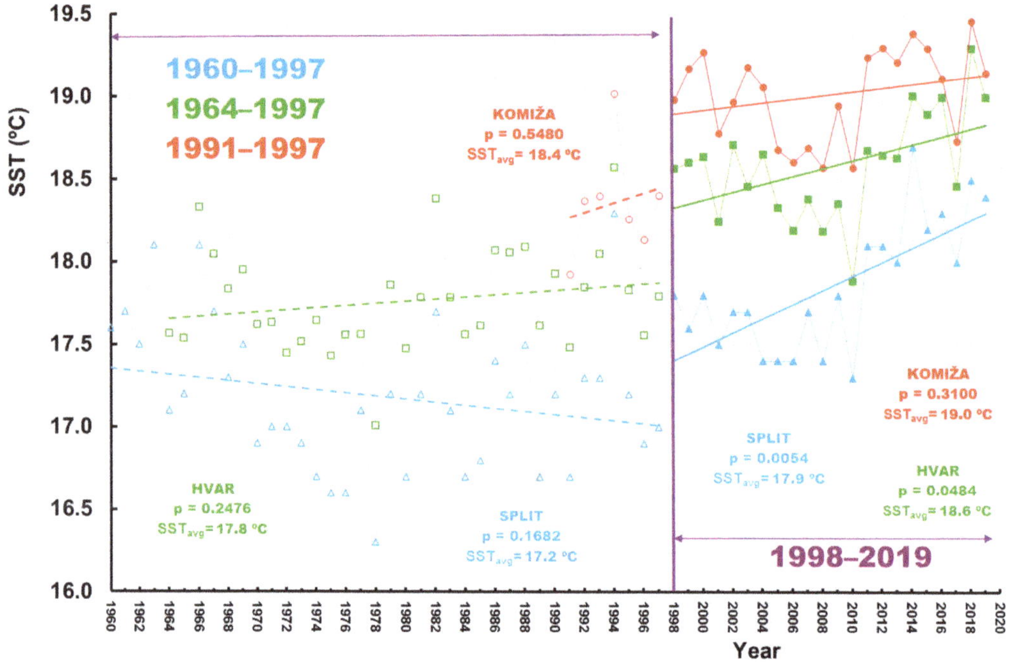

Figure 3. Time series of the mean annual SST measured at Split, Hvar, and Komiža stations. SST_{avg} is an average value of the mean annual SST, and *p* represents M–K test values, calculated for two sub–periods defined by the results of Rescaled Adjusted Partial Sums (RAPS) method.

Table 3. The average values of the mean annual SST time series within a sub–period defined by the RAPS method at the analyzed stations and the results of the F–test and the *t*-test.

Station	Sub–Period	SST_{avg} (°C)	*p* (F–test)	*p* (*t*-test)
SPLIT	1960–1997	17.19	0.565	2.8×10^{-7}
	1998–2019	17.85		
HVAR	1964–1997	17.77	0.788	7.7×10^{-13}
	1998–2019	18.58		
KOMIŽA	1991–1997	18.35	0.478	2.0×10^{-5}
	1998–2019	19.02		

Figure 4 shows the time series of the mean annual air temperatures measured at Split, Hvar, and Komiža stations. The average values of the mean annual air temperature during the analyzed periods were identical at Hvar and Komiža (16.7 °C), while in Split they were slightly lower (16.4 °C). All three stations showed statistically significant increasing trends in the mean annual air temperature (i.e., low *p*–values < 0.01). To achieve better fitting to the data, quadratic regressions were preferred over simple linear regression for the time series from all analyzed stations.

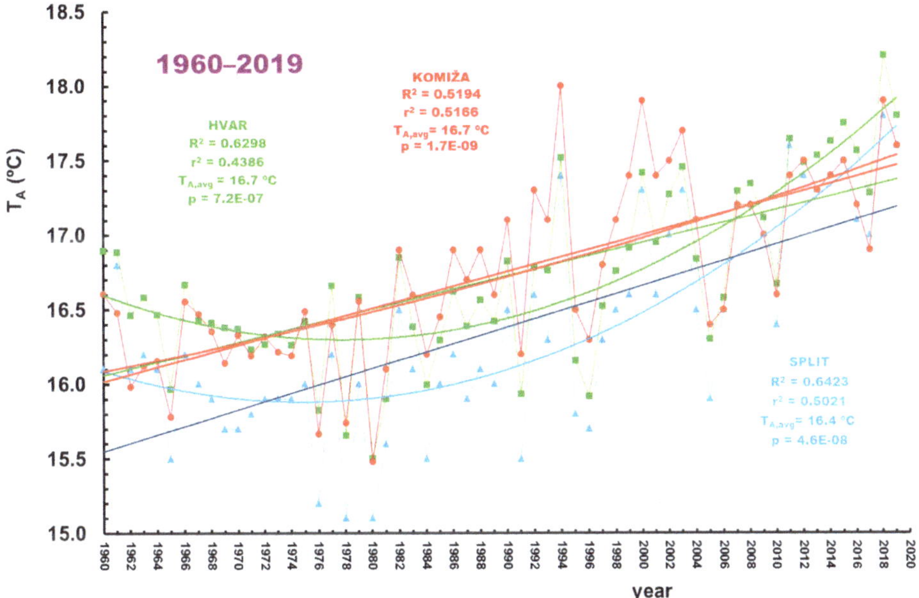

Figure 4. Time series of the mean annual air temperature measured at Split, Hvar, and Komiža stations. The r^2 and R^2 represent the square values of the correlation coefficients of the linear and quadratic regressions, respectively, and p represents M–K test values. $T_{A, avg}$ is the average value of the mean annual air temperature in the investigated period.

The RAPS method evidenced the presence of two sub–periods in the mean annual air temperature time series: (i) 1960–1991, and (ii) 1992–2019 (Figure 5). The M–K test evidenced that the trends within the first sub–period were statistically insignificant (p–values > 0.05) at all analyzed stations. However, within the second sub–period statistically significant increasing trends were observed at Hvar and Split stations ($p < 0.01$).

The average values of the mean annual air temperature within a sub–period defined by the RAPS method are shown in Figure 5 and Table 4. The statistical analyses evidenced statistically significant differences between the average values the mean annual air temperature in the two sub–periods, with the rejection of the null hypothesis of the t–test (low p–values, i.e., $p < 0.01$), and similar variances of the sub–periods reflecting the failure to reject the null hypothesis of the F–test (high p–value). The results indicated that the air temperature had started to increase considerably at the beginning of the 1990s at all analyzed stations. These results fit the regional warming patterns observed in Croatia and the western Balkans [22–25]. Furthermore, it can be concluded that the rapid increase in air temperature had occurred 6 years before the increase in SST at all analyzed stations.

The correlation between the mean annual SST and the mean annual air temperature time series was the highest in Hvar, with the values of $r^2 = 0.796$ in the period from 1964 to 2019, followed by Split with $r^2 = 0.688$ in the period from 1960 to 2019, and the lowest was in Komiža with $r^2 = 0.6183$ in the period from 1991 to 2019.

Table 5 shows the r–squared values of the linear correlation coefficients (i) between the pairs of the time series of the mean annual SST during periods of contemporaneous measurements at all three stations (from 1991 to 2019), and (ii) between the pairs of the time series of the mean annual air temperature (from 1960 to 2019).

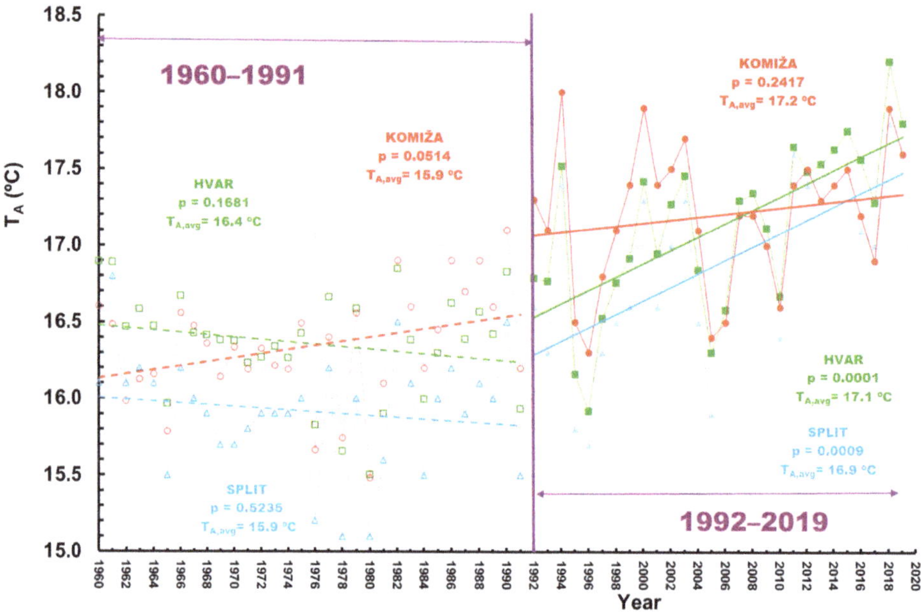

Figure 5. Time series of the mean annual air temperature measured at Split, Hvar, and Komiža stations. $T_{A,\,avg}$ is an average value of the mean annual air temperature, and p represents M–K test values, calculated for two sub–periods defined by the results of RAPS.

Table 4. The average values of the mean annual air temperature time series within a sub–period defined by the RAPS method at the analyzed stations and the results of the F–test and the t–test.

Station	Sub–Period	$T_{A,avg}$(°C)	p (F–Test)	p (t–Test)
SPLIT	1960–1991	15.98	0.664	3.4×10^{-11}
	1992–2019	17.03		
HVAR	1960–1991	16.40	0.415	1.6×10^{-10}
	1992–2019	17.26		
KOMIŽA	1960–1991	16.45	0.331	7.8×10^{-9}
	1992–2019	17.26		

Table 5. Matrix table of the r–squared values of the linear correlation coefficients, r^2, calculated from the time series of the mean annual SST and the mean annual air temperature.

Sea Surface Temperature (1991–2019)			
r^2	SPLIT	HVAR	KOMIŽA
SPLIT	1	0.869	0.751
HVAR		1	0.872
KOMIŽA			1
Air Temperature (1960–2019)			
r^2	SPLIT	HVAR	KOMIŽA
SPLIT	1	0.957	0.816
HVAR		1	0.806
KOMIŽA			1

The high values of r^2 indicated the similarity of the SST and the air temperature regimes at analyzed stations. The highest r^2 from the SST time series was observed between the closest stations, Komiža and Hvar, $r^2 = 0.87$, and the lowest between Komiža and Split, $r^2 = 0.75$. The highest r^2 value from the time series of the mean annual air temperature was observed between Split and Hvar stations, $r^2 = 0.95$, and the lowest between Hvar and Komiža, $r^2 = 0.80$.

Furthermore, the autocorrelation method was performed on the time series of the mean annual SST and air temperature measured at Split and Hvar stations, for the period from 1960 to 2019, and from 1964 to 2019, respectively. The time series from Komiža did not qualify for autocorrelation due to the insufficient duration of the time series of SST (from 1991 to 2019). The results indicated the similar behavior of the SST and the air temperature in Hvar and the air temperature in Split having a long–term autocorrelation (Figure 6). However, the autocorrelogram of the SST from the Split station is significantly different. The values of r were steady at approximately 0.5 until 6 years when a significant drop occurred. After 8 years, the values of the autocorrelation coefficient were lower than the significance threshold (0.2) meaning that the "memory of the system" was lost. Plausible causes that decreased the correlation of the time series of the mean annual SST include higher variability of SST in Split, pronounced coastal effect and local variability of climate, and bay–like topography. Considering spatially and temporally limited data measured at the station in Split, a more detailed study should be conducted to evaluate the driving force of this different behavior.

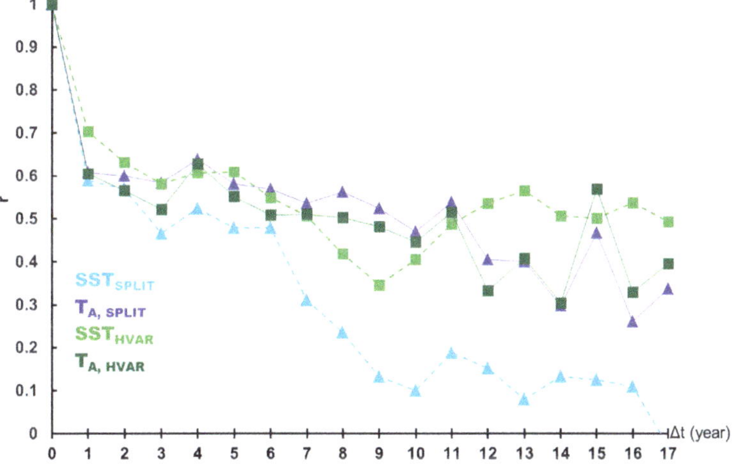

Figure 6. Autocorrelogram of the mean annual SST and the air temperature (T_A) time series measured at stations in Split (blue) and Hvar (green). Δt refers to the year in a sequence.

4.2. Analyses of the Mean Monthly Sea Surface Temperature and Air Temperature

The summary of the statistical analysis (minimum, average, maximum, and range) of the mean monthly SST and air temperature (T_A) time series, as well as their differences ($\Delta T = SST-T_A$), is shown in Table S2 of the supplementary material. The statistical analyses evidenced that the ΔT of the minimum values coincided at all three stations during the warmer period of the year (i.e., from May to August), and ranged from -1.8 to $6.4\,°C$ in Split, from -2.6 to $7\,°C$ in Hvar, and from -1.8 to $7\,°C$ in Komiža. The amplitude was slightly higher at island stations (i.e., Hvar and Komiža) than at the Split station. The ΔT of the maximum values were the lowest during the warmer period of the year (from April to September) and the highest during December at all three stations, and ranged from -4.6 to $5.3\,°C$ in Split, from -3.1 to $5\,°C$ in Hvar, and from -2.8 to $4.7\,°C$ in Komiža. The distribution of the ΔT of the average values showed a similar pattern as the

ΔT of minimum and maximum, and the values were the lowest in July and the highest in December at all analyzed stations. The ΔT of the average values ranged from −2.6 to 5 °C in Split, from −2.2 to 5.2 °C in Hvar, and from −1.6 to 5.1 °C in Komiža. The ranges of the mean monthly ΔT values were the lowest (i.e., highest negative values) during the winter period (from January to March), and the highest during warmer periods of the year at all analyzed stations. The ranges of the ΔT values were from −4.4 to −0.8 °C in Split, from −4.1 to –0.2 °C in Hvar, and from −3.7 to −0.8 °C in Komiža. At all analyzed stations, the ranges of the mean monthly air temperature were significantly higher than the ranges of the SST in each month of the year.

The average values of the mean monthly SST, air temperature, as well as their difference ($\Delta T = SST–T_A$), are shown in Figure 7.

Figure 7 shows the comparison of the average values of the mean monthly SST and air temperatures at Split, Hvar, and Komiža stations. In the warmer part of the year (from May to August), the air temperature was higher than the SST at all analyzed stations. Furthermore, the air temperature and the SST were nearly identical in April at stations in Split and Hvar. The most significant difference occurred in July when their difference was 2.64 °C in Split, 2.25 °C in Hvar, and 1.63 °C in Komiža. In the colder parts of the year, the SST was higher than the air temperature at all analyzed stations, with the highest difference in December, when the mean monthly SST was on average 5 °C higher than the mean monthly air temperature. The results evidenced a very similar behavior of temperature at all analyzed stations, despite the differences in the duration of the time series. The smallest ΔT values were observed at the station in Komiža, and the highest at the Split station. This fact could be partly explained by the differences in the duration of the time series, but also by the local effect of the position of the meteorological station and its distance from the SST measuring point. Furthermore, the position of the meteorological station in terms of the distance from the landmass also plays an important role.

Table 6 shows the slope of the linear equation, a, squared values of the linear correlation coefficient, r^2, and M–K probability values, p, for the analyzed time series of the mean monthly SST. The results indicated that the statistically significant increasing trends in SST were observed in the warmer parts of the year (March–August) in Split, throughout the year in Hvar, while the statistically more complex situation was observed in Komiža, where increasing trends occurred in March, June, July, September, and December.

Table 6. The slope of the linear equation, a, squared values of the linear correlation coefficient, r^2, and M–K probability values, p, calculated from the time series of the mean monthly SST. M–K p values $0.01 < p < 0.05$ are highlighted in blue, and $p < 0.01$ in red.

Month	Split			Hvar			Komiža		
	a	r^2	p	a	r^2	p	a	r^2	p
1	0.007	0.021	0.458	0.018	0.195	0.002	0.023	0.092	0.137
2	0.010	0.068	0.173	0.015	0.161	0.004	0.019	0.098	0.220
3	0.017	0.129	0.010	0.021	0.222	0.000	0.025	0.150	0.061
4	0.016	0.121	0.002	0.023	0.276	7.5×10^{-5}	0.053	0.468	7.5×10^{-5}
5	0.023	0.109	0.030	0.030	0.216	0.001	0.030	0.069	0.309
6	0.024	0.163	0.002	0.035	0.295	4.5×10^{-5}	0.037	0.114	0.036
7	0.024	0.217	1×10^{-4}	0.034	0.361	7.4×10^{-6}	0.040	0.220	0.016
8	0.019	0.115	0.009	0.028	0.272	0.000	0.018	0.032	0.347
9	0.014	0.047	0.071	0.021	0.114	0.009	0.035	0.080	0.034
10	0.010	0.029	0.113	0.018	0.087	0.036	0.001	2×10^{-4}	0.820
11	0.011	0.033	0.140	0.020	0.127	0.006	0.031	0.098	0.067
12	0.008	0.023	0.136	0.022	0.192	0.001	0.051	0.243	0.008

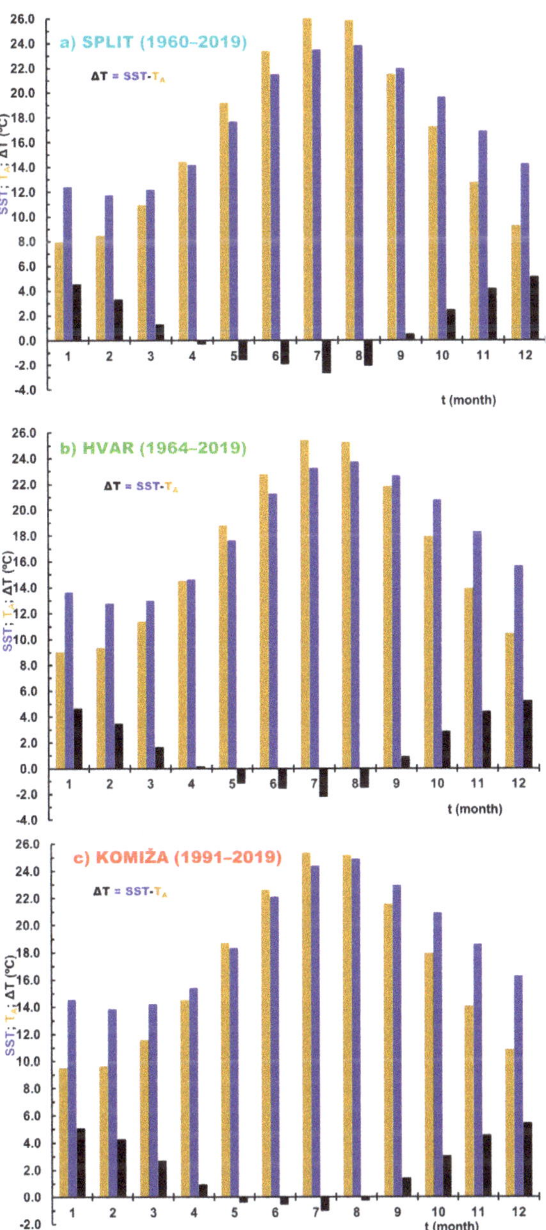

Figure 7. The average values of the monthly mean SST (blue), air temperature (brown), as well as their difference $\Delta T = SST\text{-}T_A$ (black) from the time series measured at Split (**a**), Hvar (**b**), and Komiža (**c**) stations.

Table 7 shows the slope of the linear equation, a, squared values of the linear correlation coefficient, r^2, and M–K probability values, p, calculated from the time series of the mean monthly air temperature. The results indicated the nearly identical behavior of the air temperature at all analyzed stations. Statistically significant increasing trends in the mean monthly air temperature occurred from March to August at all analyzed stations, but also in December at the Split and Komiža stations.

Table 7. The r–squared values of the linear correlation coefficient, r^2, slope of the linear equation, a, and M–K probability values, p, calculated from the time series of mean monthly air temperature. M–K p values $0.01 < p < 0.05$ are highlighted in blue, and $p < 0.01$ in red.

Month	Split			Hvar			Komiža		
	a	r^2	p	a	r^2	p	a	r^2	p
1	0.019	0.044	0.147	0.011	0.018	0.378	0.019	0.056	0.100
2	0.015	0.020	0.304	0.010	0.011	0.463	0.014	0.026	0.277
3	0.028	0.086	0.038	0.022	0.079	0.039	0.022	0.086	0.037
4	0.032	0.152	0.001	0.022	0.120	0.001	0.023	0.131	0.002
5	0.027	0.089	0.014	0.025	0.115	0.003	0.024	0.103	0.011
6	0.048	0.320	1.2×10^{-5}	0.042	0.329	8.6×10^{-8}	0.040	0.317	3.9×10^{-6}
7	0.049	0.402	9.4×10^{-7}	0.046	0.470	6.9×10^{-8}	0.043	0.395	4.1×10^{-7}
8	0.050	0.266	5.3×10^{-5}	0.044	0.336	3.1×10^{-6}	0.046	0.308	5.2×10^{-6}
9	0.012	0.020	0.255	0.016	0.046	0.097	0.015	0.042	0.109
10	0.014	0.040	0.169	0.009	0.018	0.392	0.014	0.041	0.203
11	0.019	0.053	0.103	0.014	0.031	0.184	0.021	0.075	0.050
12	0.017	0.055	0.045	0.006	0.007	0.350	0.017	0.063	0.039

Statistical analyses performed on the time series of the mean monthly SST and air temperature undoubtedly evidenced that the most significant increasing trends in the SST and the air temperature occurred during warmer parts of the year, i.e., during spring and summer. Similar warming trends could be present over the entire Adriatic Sea and its coast, but further detailed studies on the time series from the other meteorological stations, coupled with studies using gridded datasets from remote sensing or numerical simulation, are needed to assess whether these trends are present on a regional scale. Furthermore, the results of this study are in concordance with findings from Bartolini et al. [23], who conducted a regional climatological study where they analyzed air temperatures at 21 stations in the Mediterranean region, i.e., in Tuscany, Italy, and they have concluded that the most rapid and intensive warming trends occurred from March to August.

Figure 8 shows the ratio of the average values of the mean monthly SST and air temperature in the period of contemporaneous measurements at all three stations (from 1991 to 2019). The data loops from all three stations are similar in shape but are slightly shifted in values. The analyses of the time series from the period of contemporaneous measurements confirmed the previous conclusion which was based on the divergent time series.

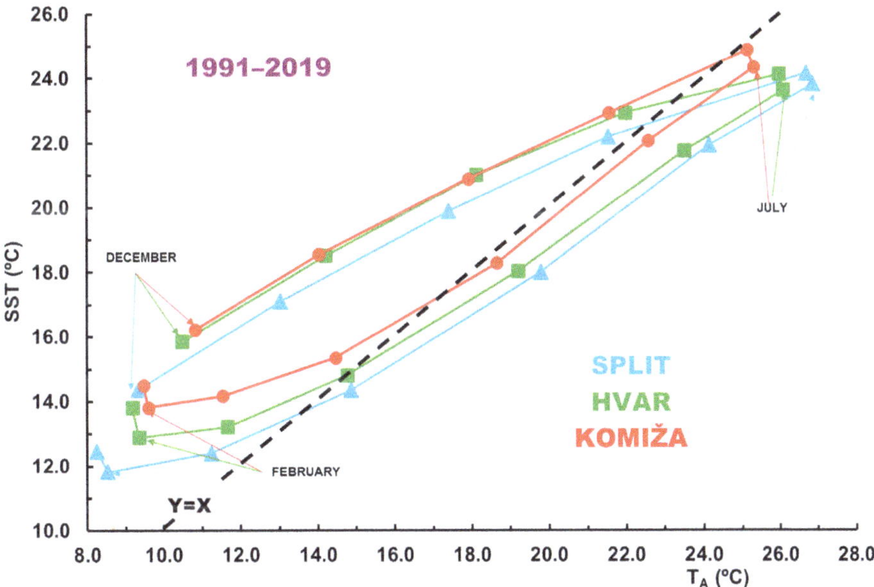

Figure 8. The ratio of the mean monthly SST and air temperature from the period of contemporaneous measurements at all three analyzed stations.

5. Conclusions

In the past 40 years, the Adriatic Sea and adjacent coastal areas have faced an increase in sea surface temperature, air temperature, and changes in the precipitation regime. Statistical analyses conducted within this study evidenced increasing trends in both the investigated temperature time series (i.e., SST and T_A). The results of RAPS indicated that the increase has been sharper since the 1990s but it occurred with a significant temporal shift (6 years) between the mean annual air temperature and SST. The observed lag in the warming of the Adriatic Sea is most likely a result of the slower response of the sea to the warming process, due to the inherent ability of the sea to absorb vast amounts of energy. Furthermore, the most significant increasing trends in the mean monthly air temperature and SST occurred during warmer parts of the year, i.e., during spring and summer. These results are in accordance with regional climate models [25,26] for the Adriatic Sea.

The climate changes described in this and other works (e.g., [22–24]) have a strong impact on the environment, the marine species, and the population living in the Adriatic region. For example, the habitats of many thermophilic species had migrated horizontally and vertically towards the deeper and the colder parts of the sea. If the sea surface temperature will continue to increase, the geographical distribution of these species will continue to decrease and will eventually cause extirpation and possibly even extinction. In addition, changes in the composition and quantity of zooplankton were observed, particularly in coastal areas of the Adriatic. Moreover, frequent blooms of marine phytoplankton and the spread of bacteria and thermophilic species of tropical algae were also observed [4].

Regional climate models for the Mediterranean region showed the continuation or even an increase in warming trends, and therefore, it is realistic to assume that the negative changes in the Adriatic Sea and its coast will be even more pronounced in the near future. In conjunction with increasing anthropogenic pressures, such as overfishing, urban and industrial pollution, the devastation of habitats, seasonal tourism pressures, and hydrocarbon exploitation, the negative consequences could be even more drastic. The lack of reliable indicators of climate change or variability, i.e., the sea surface and the air temperature, measured over a dense network of meteorological stations, has a significant influence on the development of effective measures for mitigation of negative effects of climate change [27].

Furthermore, as a result of the variable distance from the mainland, and local or regional topography, the effects of climate change manifest differently in specific islands or coastal regions. Limited and vulnerable groundwater resources along the coast of the Croatian part of the Adriatic Sea and in the related islands, in combination with unsustainable anthropic activities (e.g., mass tourism, land–use changes, groundwater over–abstraction, and urbanization), significantly reduce the options for adaption to current and future climate change. Due to the vast cultural, historical, social, geographical, and biological diversity, the Mediterranean region requires urgent and effective measures that will foster its sustainable development. The fundamental problem is that the Croatian part of the Adriatic Sea, similar to the other countries in the Mediterranean region, does not have a sufficiently dense network of meteorological stations and sufficiently long time series of measured data, especially sea surface temperature data.

The availability of high–resolution data on climate change and variability could enable island communities to enhance their resilience and design site–specific measures to mitigate possible negative effects on water availability and natural ecosystems. Besides the structural modifications (e.g., re–use of purified domestic and industrial wastewater, reduction in losses from water supply systems, desalinization plants, managed aquifer recharge), a holistic approach could also be fostered by increasing awareness and education of the local population on correct utilization of the water resource (e.g., promotion of water savings during dry months, rainwater harvesting, planting of crops that require little or no irrigation, reduction in carbon footprint, and preservation of ecosystems and their services to mitigate floods and droughts).

The authors hope that this study will contribute to a better understanding of this topic and that it will initiate interdisciplinary cooperation and discussion on the more intensive and coordinated investigation of this complex and exceedingly important issue for Croatia, the Adriatic, and the Mediterranean region.

Supplementary Materials: The following are available online at https://www.mdpi.com/article/10.3390/jmse9040358/s1. The following Supplementary Materials are submitted alongside the manuscript: Figure S1a. The RAPS visualization of the mean annual SST time series; Figure S1b. The RAPS visualization of the mean annual air temperature time series; Table S2. Statistics (minimum, average, maximum, and range) of the mean monthly SST, air temperature (TA), and their difference ($\Delta T = SST\text{-}T_A$) from the time series measured at Split, Hvar, and Komiža stations. The negative values of ΔT were highlighted in red.

Author Contributions: Conceptualization, investigation, and writing of original draft, O.B.; investigation, visualization, and data curation, D.B.; data curation and validation, review and editing, M.P. (Matko Patekar); supervision, writing and editing, and validation, M.P. (Marco Pola). All authors have read and agreed to the published version of the manuscript.

Funding: This research was supported by the Croatian Geological Survey, Department of Hydrogeology and Engineering Geology.

Institutional Review Board Statement: Not applicable.

Informed Consent Statement: Not applicable.

Data Availability Statement: The data used within this study are the property of the Croatian Meteorological and Hydrological Service (DHMZ). Terms of use, data availability, and contact can be found at: https://klima.hr/razno/katalog_i_cjenikDHMZ.pdf.

Acknowledgments: Data used in this study was provided by courtesy of the Croatian Meteorological and Hydrological Service, for which we thank them.

Conflicts of Interest: The authors declare no conflict of interest.

References

1. Viličić, D. Jadran i globalne promjene [The Adriatic and the global changes]. *Priroda* **2013**, *13*, 22–28.
2. Viličić, D. Specifična oceanološka svojstva hrvatskog dijela Jadrana [Specific oceanographic characteristics of the Croatian part of The Adriatic Sea]. *Hrvatske Vode* **2013**, *22*, 297–314.
3. Grbec, B.; Morović, M.; Matić, F.; Ninčević Gladan, Ž.; Marasović, I.; Vidjak, O.; Bojanić, N.; Čikeš Keč, V.; Zorica, B.; Kušpilić, G.; et al. Climate regime shifts and multi-decadal variability of the Adriatic Sea pelagic ecosystem. *Acta Adriatica* **2015**, *56*, 47–66.
4. Krželj, M. Utjecaj klimatskih promjena na morski okoliš [Climate change effects on the marine environment]. *Paediatr. Croat.* **2010**, *54*, 18–23.
5. Orlić, M. Croatian coastal waters. In *Physical Oceanography of the Adriatic Sea: Past, Present and Future*; Cushman-Roisin, B., Gačić, M., Poulain, P.-M., Artegiani, A., Eds.; Kluwer Academic Publishers: Dordrecht, The Netherlands, 2001; pp. 189–214.
6. Bonacci, O.; Ljubenkov, I. Različite vrijednosti i trendovi temperatura zraka na dvije postaje na malom otoku: Slučaj meteoroloških postaja Korčula i Vela Luke na otoku Korčuli [Different values and air temperature trends at meteorological stations on small island of Korčula: Case study of meteorological stations in Vela Luka and Korčula]. *Hrvat. Vode* **2019**, *28*, 183–196.
7. Bonacci, O.; Patekar, M.; Pola, M.; Roje-Bonacci, T. Analyses of climate variations at four meteorological stations on remote islands in the Croatian part of the Adriatic Sea. *Atmosphere* **2020**, *11*, 1044. [CrossRef]
8. Vlahakis, G.N.; Pollatou, R.S. Temporal variability and spatial distribution of Sea Surface Temperatures in the Aegean Sea. *Theor. Appl. Climatol.* **1993**, *47*, 15–23. [CrossRef]
9. Bulgin, C.E.; Merchant, C.J.; Ferreira, D. Tendencies, variability and persistence of sea surface temperature anomalies. *Sci. Rep.* **2020**, *10*, 7986. [CrossRef] [PubMed]
10. Penzar, B.; Penzar, I.; Orlić, M. *Vrijeme i Klima Hrvatskog Jadrana (Weather and Climate of the Croatian Adriatics)*, 1st ed.; Dr. Frletar: Zagreb, Croatia, 2001.
11. Duplančić Leder, T.; Ujević, T.; Čala, M. Coastline lengths and areas of islands in the Croatian part of the Adriatic Sea determined from the topographic maps at the scale 1:25,000. *Geoadria* **2004**, *9*, 5–32.
12. Šegota, T.; Filipčić, A. Köppenova podjela klima i hrvatsko nazivlje [Köppen climate classification and the Croatian nomenclature]. *Geoadria* **2003**, *8*, 17–37. [CrossRef]
13. Pandžić, K. *Naputak za Opažanja i Mjerenja na Glavnim Meteorološkim Postajama (Instructions for Observation and Measurements at Main Meteorological Stations)*; Državni hidrometeorološki zavod: Zagreb, Croatia, 2008.
14. Heino, R. Climate in Finland during the period of meteorological observations. *Finn. Meteorol. Inst. Contrib.* **1994**, *12*, 209.
15. Bonacci, O.; Željković, I.; Trogrlić, R.Š.; Milković, J. Differences between true mean daily, monthly and annual air temperatures and air temperatures calculated with three equations: A case study from three Croatian stations. *Theor. Appl. Climatol.* **2013**, *114*, 271–279. [CrossRef]
16. Gough, W.A.; Žaknić-Ćatović, A.; Zajch, A. Sampling frequency of climate data for the determination of daily temperature and daily temperature extrema. *Int. J. Climatol.* **2020**, *40*, 5451–5463. [CrossRef]
17. Husain Shourov, M.M.; Mahmud, I. pyMannKendall: A python package for non parametric Mann Kendall family of trend tests. *J. Open Source Softw.* **2019**, *4*, 1556. [CrossRef]
18. Garbrecht, J.; Fernandez, G.P. Visualization of trends and fluctuations in climatic records. *Water Resour. Bull.* **1994**, *30*, 297–306. [CrossRef]
19. Bonacci, O.; Roje-Bonacci, T. Primjena metode dan za danom (day to day) varijabilnosti temperature zraka na podatcima opaženim na opservatoriju Zagreb-Grič (1887–2018). *Hrvat. Vode* **2019**, *28*, 125–134.
20. McGhee, J.W. *Introductory Statistics*; West Publishing Company: Saint Paul, MN, USA, 1985.
21. Mangin, A. Pour une meilleure connaissance des systemes hydrologiques a partir des analyses correlatioire et spectrale. *J. Hydrol.* **1984**, *67*, 25–43. [CrossRef]
22. Bonacci, O. Analiza nizova srednjih godišnjih temperature zraka u Hrvatskoj. *Građevinar* **2010**, *62*, 781–791.
23. Bonacci, O. Increase of mean annual surface air temperature in the Western Balkans during last 30 years. *Vodoprivreda* **2012**, *44*, 75–89.
24. Bartolini, G.; di Stefano, V.; Maracchi, G.; Orlandini, S. Mediterranean warming is especially due to summer season–Evidences from Tuscany (central Italy). *Theor. Appl. Climatol.* **2012**, *107*, 279–295. [CrossRef]
25. Branković, Č.; Güttler, I.; Gajić-Čapka, M. Evaluating climate change at the Croatian Adriatic from observations and regional climate models' simulations. *Clim. Dyn.* **2013**, *41*, 2353–2373. [CrossRef]
26. Pandžić, K.; Kobold, M.; Oskoruš, D.; Biondić, B.; Biondić, R.; Bonacci, O.; Likso, T.; Curić, O. Standard normal homogeneity test as a tool to detect change points in climate-related river discharge variation: Case study of the Kupa River Basin. *Hydrol. Sci. J.* **2020**, *65*, 227–241. [CrossRef]
27. Al Sayah, M.J.; Abdallah, C.; Khouri, M.; Nedjai, R.; Darwich, T. A framework for climate change assessment in Mediterranean data-sparse watersheds using remote sensing and ARIMA modeling. *Theor. Appl. Climatol.* **2020**, *143*. [CrossRef]

Journal of
*Marine Science
and Engineering*

Article

Anomalous Oceanic Conditions in the Central and Eastern North Pacific Ocean during the 2014 Hurricane Season and Relationships to Three Major Hurricanes

Victoria L. Ford [1,*,†], Nan D. Walker [1] and Iam-Fei Pun [2]

[1] Department of Oceanography and Coastal Sciences, Coastal Studies Institute Earth Scan Laboratory, Louisiana State University, Baton Rouge, LA 70803, USA

[2] Graduate Institute of Hydrological and Oceanic Sciences, National Central University, Taoyuan 320, Taiwan

*   Correspondence: victoriaford@tamu.edu

†   Current institution: Climate Science Lab, Department of Geography, Texas A&M University, College Station, TX 77845, USA.

Received: 27 February 2020; Accepted: 14 April 2020; Published: 17 April 2020

Abstract: The 2014 Northeast Pacific hurricane season was highly active, with above-average intensity and frequency events, and a rare landfalling Hawaiian hurricane. We show that the anomalous northern extent of sea surface temperatures and anomalous vertical extent of upper ocean heat content above 26 °C throughout the Northeast and Central Pacific Ocean may have influenced three long-lived tropical cyclones in July and August. Using a variety of satellite-observed and -derived products, we assess genesis conditions, along-track intensity, and basin-wide anomalous upper ocean heat content during Hurricanes Genevieve, Iselle, and Julio. The anomalously northern surface position of the 26 °C isotherm beyond 30° N to the north and east of the Hawaiian Islands in 2014 created very high sea surface temperatures throughout much of the Central Pacific. Analysis of basin-wide mean conditions confirm higher-than-average storm activity during strong positive oceanic thermal anomalies. Positive anomalies of 15–50 kJ cm^{-2} in the along-track upper ocean heat content for these three storms were observed during the intensification phase prior to peak intensity, advocating for greater understanding of the ocean thermal profile during tropical cyclone genesis and development.

Keywords: sea surface temperature; upper ocean heat content; hurricane intensity; Northeast Pacific; Hawaii; Hurricane Genevieve; Hurricane Iselle; Hurricane Julio

1. Introduction

Due to the potentially destructive and deadly nature of powerful tropical cyclones (TCs), a more complete understanding of their genesis and life cycles is imperative. Of the many associated risks, the greatest threats to public health and life are storm surges, flooding, tornadoes, torrential rainfall, and high winds [1]. The 2014 hurricane season in the Northeast Pacific Ocean (NPO) basin was more active than normal. Of the 22 named storms that formed, 16 became hurricanes, and 9 attained major hurricane status (Category 3 and higher; Figure 1). We specifically include Central Pacific TCs in these totals. Long-term averages for the NPO (1981–2010) include 15 tropical storms, 8 hurricanes, and 4 major storms [2]. The NPO accumulated cyclone energy, an index of storm occurrence, duration, and intensity [3], was 162% of the long-term median value [2], indicating that the 2014 NPO season was one of the busiest on record. As a direct result of the 2014 NPO hurricane season, 27 lives were lost and over USD $1.02 billion in damages occurred [2].

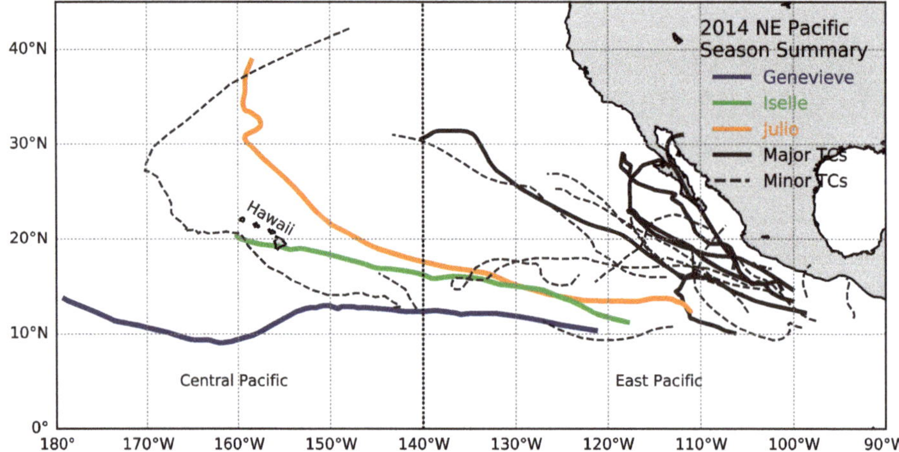

Figure 1. Tropical cyclone (TC) tracks of the 2014 Northeast Pacific hurricane season. Minor TCs (defined as Category 2 status and below) are noted by thin grey dashed line, while major TCs (Category 3 status and above) are in thick grey. TCs of particular interest to our study are highlighted in color—Hurricane Genevieve in blue, Hurricane Iselle in green, and Hurricane Julio in orange. The boundary between the East and Central Pacific basins at 140° W is shown with a vertical gray line.

Of particular interest were the three major TCs that approached the Hawaiian Islands, the third highest occurrence since 1949 [4]. Major hurricane Iselle made a rare landfall on the Island of Hawaii in August 2014 as a strong tropical storm, the first landfall reported there since 1871 [5]. Berg and Kimberlain [6] reported that the Big Island of Hawaii, experienced up to an eight-foot storm surge resulting in week-long power outages and extensive damages to agricultural crops amounting to more than USD $50 million.

TC genesis is influenced by several key atmospheric and oceanic factors; namely, temperature at the surface and extending through sufficient depth in the upper ocean, an initial atmospheric disturbance, a vertical humidity gradient in the lower atmosphere, sufficient Coriolis force, and low vertical wind shear. We specifically focus on the oceanic component of high oceanic temperatures at the sea surface and in the upper ocean involved in generation and maintenance of three NPO TCs in this paper. The required minimum surface temperature to support cyclogenesis is widely agreed to be 26–27 °C [7]. DeMaria and Kaplan [8] and Whitney and Hobgood [9] found that sea surface temperature (SST) acts as an upper limit on the maximum potential intensity of a TC but indicated that other environmental factors also influence the actual intensity of any given TC. Recent oceanic warming has increased tropical SSTs by approximately 0.5 °C between 1970 to 2004, and coincident increases in the number of TCs reaching Category 4 status and higher in the NPO [10].

Perhaps of even greater importance than high SSTs is a sufficiently deep upper ocean thermal structure, which is theorized to energize and intensify tropical cyclones. Leipper and Volgenau [7] coined 'hurricane heat potential' as the excess of heat above 26 °C within the upper ocean. In the tropics, SSTs regularly exceed 26 °C, and thus the depth to which the 26 °C isotherm extends is also vital. Upper ocean heat content (UOHC) and hurricane heat potential [11] are used interchangeably in the literature. For temperatures above this threshold, UOHC provides a vertically integrated estimate of available energy within the upper ocean [12]. The importance of UOHC on intensity changes has been increasingly discussed worldwide (e.g., in the Gulf of Mexico [13]; the North Atlantic Ocean [14]; and the NPO [15]). Lin et al. [16] stressed the criticality of the ocean thermal structure in controlling intensity, especially for the most intense Western NPO storms and those that move slowly. Research on the upper ocean response to a moving TC has also focused on the upwelling and entrainment of cool, deep waters which lead to a reduction in enthalpy fluxes, the primary energy source for TC maintenance (e.g., [17–20]).

In this study, we analyze the 2014 oceanic conditions during an active hurricane season in the context of sea surface and sub-surface warming across the last decade 2009–2019 in the Eastern and Central NPO. We employ microwave SST observations, UOHC calculations, and their respective anomalies and hypothesize that in addition to high SSTs at genesis, the surface location of the 26 °C isotherm is important for diagnosing the basin-wide surface ocean conditions, as well as maintenance of a high UOHC anomaly. The position of the 26 °C isotherm, both at the surface and at depth, is also believed to be of particular interest as a potential driver of the major hurricane activity in the Central NPO and near the Hawaiian Islands, including a rare landfall on the south coast of the Island of Hawaii.

2. Materials and Methods

2.1. Sea Surface Temperature and Anomalies

This study utilized a wealth of remotely sensed satellite measurements to identify upper ocean thermal structure and anomalous oceanic conditions. The gridded Microwave Optimally Interpolated Sea Surface Temperature version 4 data were obtained from Remote Sensing Systems (REMSS). This dataset blends all data within one day of retrieval from several satellite platforms that have lower frequency channels (6–7 GHz and/or 11 GHz): the Advanced Microwave Sounding Radiometer 2 (AMSR-2), the WindSat radiometer, and the Tropical Rainfall Measuring Mission/Microwave Imager (TRMM/TMI) [21]. Microwave radiometers are especially well suited for observing the sea surface in tropical cyclone environments as the ability to measure SST through clouds is not impeded. Spatial and temporal resolutions are 0.25° and daily, respectively, with an SST measurement accuracy of approximately 0.5 °C [22]. To identify cool wake contamination bias in the SST on TC approach to any particular location, SST values were extracted along-track at daily intervals of 0 to 4 days prior to actual time of hurricane passage [23]. SST '4 days prior' was deemed the most representative of uncontaminated ocean conditions and is used throughout this study (Figure S1). Further notes on SST product quality control are available through REMSS.

National Oceanic and Atmospheric Administration (NOAA) high-resolution SST anomaly version 2 (SSTA) data were provided by the Earth System Research Laboratories Physical Sciences Department (ESRL PSD), which is a high-resolution optimally interpolated blended analysis of SSTs and sea ice concentration from the infrared sensor Advanced Very High Resolution Radiometer (AVHRR) and in-situ ship and buoy observations [24]. Gridded data are available at 0.25° spatial and daily temporal resolutions. The SST monthly anomaly values represent daily observed SST deviations from the long-term 30-year climatological mean covering 1971 to 2000 [25] and highlight the abnormally high ocean temperatures throughout the NPO in our study. We specifically note that the SST and SSTA products used in this study were derived from different remote sensing sensors, each with their own respective inherent biases (see [21,24,25]), and thus exercise caution in discussing the relationship between SST and SSTA products.

2.2. Upper Ocean Heat Content and Anomalies

Upper ocean heat content is a derived variable calculated from satellite microwave SST measurements and contemporaneous satellite altimetry observations. Data for the NPO were derived from a regression model developed first for the Western North Pacific Ocean [26]. The regression method for the Western North Pacific (REGWNP) was developed to improve upon the vertical resolution of the two-layer reduced gravity model (TLM) used in many North Atlantic and NPO oceanic thermal structure studies (e.g., [11,13,15,26]). The TLM calculates upper ocean layer thickness based on the depth of the 20 °C isotherm, in-situ temperature profiles, and AVHRR satellite-retrieved SST [27]. In the REGWNP method, a temperature difference is first calculated between in-situ Argo surface and subsurface temperature profiles, REMSS microwave SST, and the World Ocean Atlas

2001 climatological temperature profiles. The isotherm displacement is linearly regressed onto the corresponding sea surface height anomaly (SSHA) by least squares fit, and varies by location, isotherm, and month. The process is repeated for isotherm depths between 4 and 29 °C. Provided with daily microwave SST, SSHA, and monthly climatological mixed layer depth from the U.S. Naval Research Laboratory, the daily oceanic thermal structure was computed. UOHC was then derived from the oceanic thermal structure profile for the depth of the 26 °C isotherm. The resulting gridded dataset has a spatial resolution of 0.25° and daily temporal resolution. Accuracy estimates of the REGWNP UOHC derivation are discussed in Pun et al. [26]. It is important to note that heat content is a relative quantity where any sub-surface isotherm could be used as a reference. This study used 26 °C as the reference isotherm in determining UOHC values in the NPO. The along-track depths of the 26 °C isotherm during Hurricanes Genevieve, Iselle, and Julio are provided for reference in Figure S2. UOHC anomaly (UOHCA) is the daily UOHC departure [26,28] from the long-term World Ocean Atlas 2013 monthly mean climatological profile covering 1955–2012.

2.3. Tropical Cyclone Observations

We made use of the National Hurricane Center's Hurricane Database 2 (HURDAT2), a best-track post-storm quality-controlled observational record database for the North Atlantic and NPO. Observations from all TCs that attained major hurricane status (wind speeds greater than 50 ms^{-1}) in the 2014 Northeast/North Central Pacific (NE/NC) version of the HURDAT2 were analyzed in Ford [23]. Uncertainty estimates for the North Atlantic best-track parameters are available, and may be applicable to the NPO as well [29]. Latitude, longitude, and maximum wind speed from the NE/NC HURDAT2, as well as SST and UOHC, were spline interpolated to create an hourly observation record for Hurricanes Genevieve, Iselle, and Julio. The interpolation process for the NPO was adapted from the North Atlantic interpolation method found in Elsner and Jagger [30], which preserves values at regular 6 h time intervals, and a piecewise polynomial for values in-between times. For TC location, spline interpolation was determined by spherical geometry. It is noted that the hourly-interpolated data used within are derived, and thus, not best-track quality controlled.

2.4. Methods

Along-track plots of SST, SSTA, UOHC, UOHCA, and corresponding maximum wind speeds (U_Z; ms^{-1}) were investigated. We compared the hourly-interpolated data, the NE/NC HURDAT2 database, and the SST and UOHC datasets for Hurricanes Genevieve, Iselle, and Julio each hour of their duration. Monthly averages of SST, UOHC, and their anomalies illustrate the general oceanic conditions during this time. In this analysis, the 26 °C isotherm was used as the reference threshold from which UOHC was calculated [7].

3. Results

3.1. General Oceanic Conditions

To investigate variability of ocean conditions throughout the NPO in 2014, we present monthly averages of SST, SSTA, UOHC, and UOHCA conditions for July and August as these months included Hurricanes Genevieve, Iselle, and Julio. These three major hurricanes are of specific interest, firstly because they initially formed within an 11-day period at the end of July into early August, and because these three TCs experienced significantly long durations and trajectories into the Central portion of the NPO. Hurricane Hernan, which also formed within the 11-day period, is not discussed here as it was a short-lived minor hurricane that remained within the Eastern NPO.

In July, SST was spatially favorable for TC formation, in excess of 26 °C from the equator to approximately 20–25° N (east of 140° W) and to roughly 30° N (west of 140° W) (Figure 2A). The location of the 26 °C isotherm at the surface is indicated in Figure 2A with a blue dashed line to depict the main area where SST was conducive for TC development and maintenance. In terms of elevated SST, regions of interest included the western Mexico coastline where SSTs were observed above 30 °C and to the northwest of the Hawaiian Islands, where SSTs in excess of 26 °C extended as far north as 30° N. Hurricanes Genevieve and Iselle (genesis locations shown by star in Figure 2) formed in July in regions of relatively high SSTs of 28.95 °C and 28.65 °C, respectively. Positive SSTAs were widespread across the study area (Figure 2B) and ranged from 0–2 °C throughout the TC generation area, defined as the longitudinal band between 10° to 20° N in the NPO tropics [31]. The maximum observed SSTAs in July were located in the Eastern NPO and exceeded 2 °C across a large region north of 15° N and south of 5° N. (Figure 2B). Iselle's westward trajectory followed the surface location of the 26 °C isotherm closely between 135° and 155° W before striking the Big Island of Hawaii. Genevieve travelled a more southerly path, but also experienced mostly positive SSTAs along its track. A region of weak negative anomaly (0 to −0.5 °C) was observed north of 20° N between 125° and 140° W, and a stronger negative anomaly of up to −2 °C was detected near 40° N near the International Date Line. Between those negative SSTA regions, a strong positive SSTA zonal-oriented band extended from 130° W to 180°, and may be related to the "Blob" generated during the 2013–2014 winter from abnormally low oceanic heat loss, due to abnormally high sea level pressure [32]. Although the "Blob" was centered in the extratropics, it extended southward to 30° N between 125°–150° W. To summarize, the dominant observation regarding SST in July 2014 was that of an anomalously high surface temperature regime throughout the Central and Eastern NPO where TCs formed and tracked during hurricane season.

UOHC in July averaged between 25–35 kJ cm^{-2} across much of the NPO in areas where SSTs exceeded 26 °C (Figure 2C), with regions of higher values centered off the coast of Acapulco, Mexico and extending offshore to the region where Iselle and Genevieve formed. The highest UOHC detected emanated from the Western Pacific Warm Pool into the NPO as far east as 135° W between the equator and 7° N, south of the normal TC generation region. UOHC in this region exceeded 150 kJ cm^{-2}, indicating very high upper ocean temperatures that reached a considerable depth. Although UOHC values northwest of the Hawaiian Islands extended beyond 30° N, values were higher west and southwest of the Islands. At the time of genesis, Genevieve and Iselle experienced 46 kJ cm^{-2} and 61.5 kJ cm^{-2} of UOHC, respectively, well above the basin-wide average. We note that Genevieve made a southerly turn at approximately 155° W towards the higher UOHC tropical region before transitioning into a major hurricane. In Figure 2D, the UOHC anomaly reveals that much of the NPO tropical region experienced a positive UOHC anomaly between 0 to 40 kJ cm^{-2}. The high UOHC region close to the western tropics exceeded 80 kJ cm^{-2} in UOHC anomaly. Although Iselle formed in an area of positive UOHCA, it tracked over lower values of UOHC and near 0 kJ cm^{-2} UOHCA after it reached 125° W. It even tracked over small regions of negative UOHCA (Figure 2C,D). North of the Hawaiian Islands we observed evidence of positive UOHC anomalies ranging from 0 to 10 kJ cm^{-2}, primarily due to the surface expression of the 26 °C isotherm above 30° N.

By August, the surface location of the 26 °C isotherm extended well above the Hawaiian Islands approaching 32–33° N between 150° and 180° W (Figure 3A). Notable around Iselle's trajectory was the northward shift of the 26 °C isotherm at the surface between 140° and 150° W towards 18° N, which was approximately parallel to the southern coastline of the Big Island of Hawaii. SSTs surrounding the Baja Peninsula and western Mexico's coastline increased in value, while the 26 °C isotherm at the surface in the eastern tropics also shifted northward approximately 1°. Hurricane Julio followed close behind Iselle, also following along the surface expression of the 26 °C isotherm, especially to the northeast and north of the Hawaiian Islands. It is notable that Julio transitioned from hurricane status to a tropical storm (thick line to dashed line; Figure 3A) around the terminus of the 26 °C isotherm's northern extent. SST anomaly in August revealed a strengthening of anomalous temperatures in the NPO, especially in a widespread region surrounding the Baja Peninsula and extending southwestward to approximately 140° W. Basin-wide, SST anomalies averaged between 1.0 and 1.5 °C. The high SST anomaly region converging around the Baja Peninsula, coupled with very high SSTs above 30 °C, provided extremely favorable long-term conditions for TC development close to the coast. The high SST anomaly region associated with the "Blob" north of 30° N covered a large swath of ocean between 130° and 150° W with positive anomalies higher than 2.0 °C, but below 26 °C SST value, which was why there was no indication of the "Blob" in our presently defined UOHC/UOHCA record. While not conducive for TC development and maintenance, anomalously high SSTs in this extratropical area had important fishery implications and regional weather impacts for the western U.S. coastline [32].

An inspection of August UOHC and UOHCA revealed a continuation of an anomalous spatial distribution of the warm surface layer, especially in the Central NPO to the west, north, and south of the Hawaiian Islands (Figure 3C,D). While UOHC values above 20° N averaged around 20 kJ cm^{-2}, the connection to the wide swath of SST anomaly in the extratropics makes a compelling case for a 'warmer-than-average' diagnosis for the Central NPO in 2014 from the equator to 30° N. A smaller, confined loci of high UOHC along the Mexican coastline moved to the southwest compared to July, influencing the area of Julio's generation. South of 10° N, the eastward extension of the West Pacific Warm Pool produced a long tongue-like feature of high UOHC, a feature that was also revealed in the UOHCA patterns (Figure 3C,D). UOHC values approached 200 kJ cm^{-2} within this southerly feature that may have impacted the intensification of Genevieve, as a result of the anomalously deep warm layer. Positive UOHC anomalies (Figure 3D) were observed throughout much of the NPO ranging from 5 to 60 kJ cm^{-2}, with two main regions of higher UOHC anomaly: off the western Mexico coastline and the western extension of high UOHC anomaly in the Central NPO. These two regions each exceeded 70 kJ cm^{-2}. Notable is a region of negative UOHCA at the equator near 146° W of less than −10 kJ cm^{-2} that spatially aligned with a similar pattern in SST and SST anomaly. It is of great interest that although Iselle and Julio formed in areas of high SST, SSTA, UOHC, and UOHCA, they were able to maintain hurricane strength over waters with minimal UOHC and low SST.

To determine how unusual the observed SST and UOHC anomalies were in July and August 2014, we calculated the monthly mean SST and UOHC anomalies across the zonal latitude band between 10–20° N in the NPO over the most recent decade 2009–2019 (Figure 4). The vast majority of TC cyclogenesis occurs in this band for the Eastern NPO [31] and Central NPO [23]. We also show TC counts in July and August for the NPO, broken down by basin—East NPO or Central NPO cyclogenesis locations—and by intensity for those storms above major hurricane status (Category 3 and above on the Saffir–Simpson scale) (Figure 4A). In 2009, the number of July and August TCs in the NPO exceeded 12 storms, with 2 additional TCs originating in the Central NPO. Following this period, TC counts in July and August remained below 6 storms until 2013 when TC counts increased again for a total of 11 storms. In 2014, there were 10 TCs originating in the Eastern NPO, 4 of which attained major hurricane status. A star is denoted in 2014 for Hurricane Iselle, which made a rare landfall on Hawaii. The number of storms increased in July and August of 2015 and 2016. Hurricane Darby made landfall in 2016 on the Island of Hawaii, the second landfalling strike in two years. Following 2016, TC counts in the NPO decreased slightly, ranging between 9 and 11 storms.

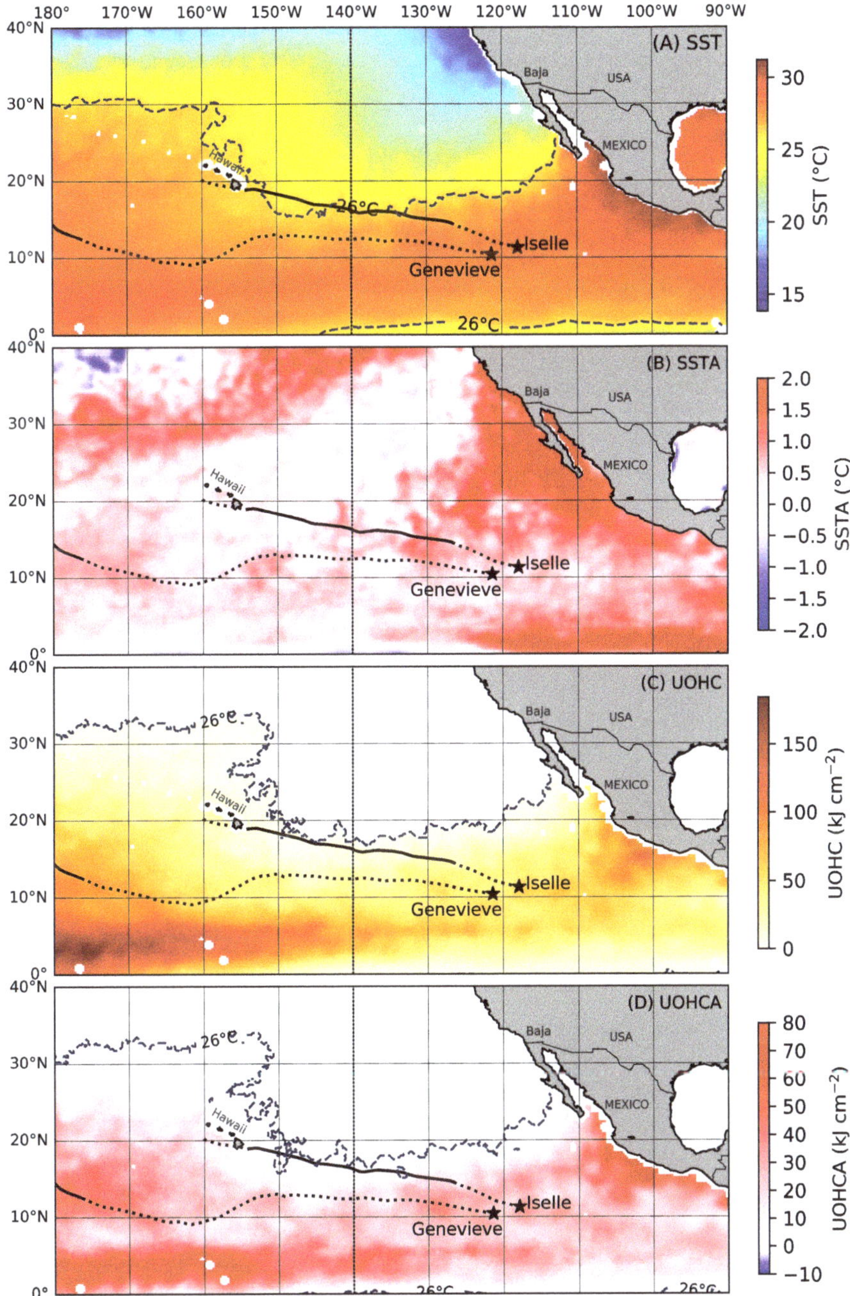

Figure 2. Oceanic conditions and anomalies during July 2014 in the Northeast Pacific Ocean (NPO) as seen in (**A**) sea surface temperature, (**B**) sea surface temperature anomaly, (**C**) upper ocean heat content, and (**D**) upper ocean heat content anomaly. The surface location of the 26 °C isotherm is contoured in blue for reference. The trajectories of Hurricanes Genevieve (25 July–7 August) and Iselle (31 July–9 August) are provided, with a star representing the genesis location. TC trajectory denoted by: black dashed line for tropical storm status (below 33 ms^{-1}) and thick black line for hurricane status (above 33 ms^{-1}).

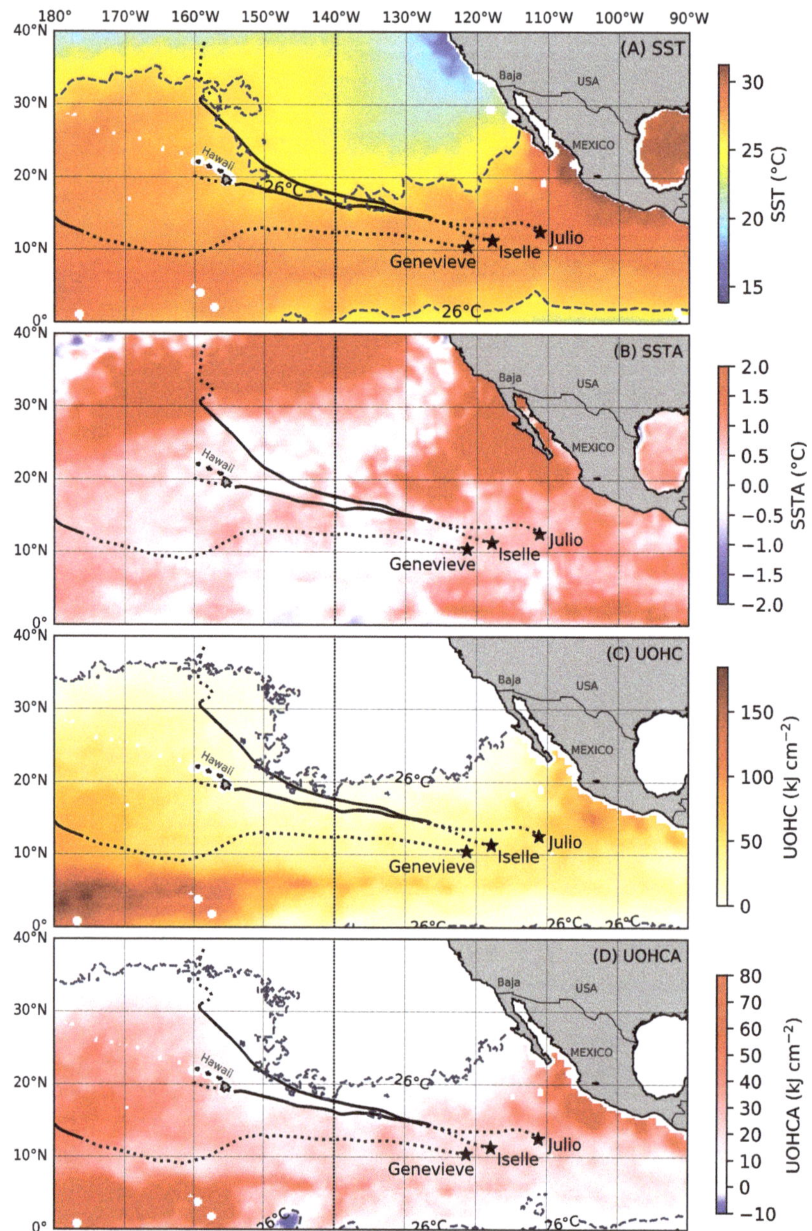

Figure 3. Oceanic conditions and anomalies during August 2014 in the Northeast Pacific Ocean (NPO) as seen in (**A**) sea surface temperature, (**B**) sea surface temperature anomaly, (**C**) upper ocean heat content, and (**D**) upper ocean heat content anomaly. The surface location of the 26°C isotherm is contoured in blue for reference. The trajectories of Hurricanes Genevieve (25 July–7 August), Iselle (31 July–9 August), and Julio (2–18 August) are provided, with a star representing the genesis location. TC trajectory denoted by: black dashed line for tropical storm status (below 33 ms^{-1}) and thick black line for hurricane status (above 33 ms^{-1}).

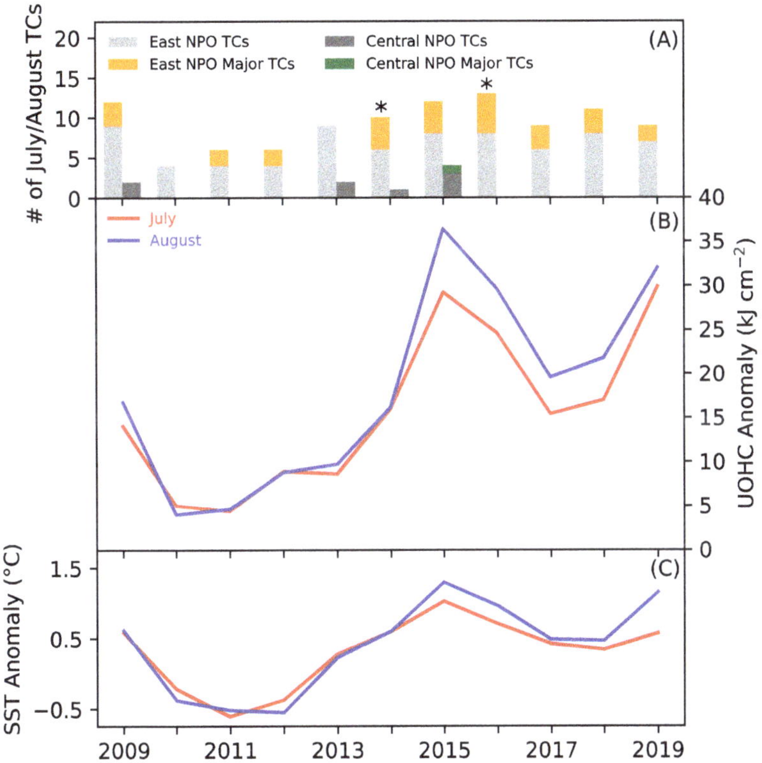

Figure 4. (**A**) July and August TC counts for 2009–2019, broken down by genesis in either the East NPO or the Central NPO, and by minor or major hurricane status. Stars indicate a Hawaiian landfalling TC in this timeframe. July and August monthly mean (**B**) upper ocean heat content (UOHC) anomalies and (**C**) sea surface temperature (SST) anomalies in the 10–20° N NPO cyclogenesis region 2009–2019.

Mean UOHC anomalies in 2009–2019 indicate positive anomalies throughout the cyclogenesis region, ranging from 5–10 kJ cm^{-2} between 2010 and 2013, and as high as 36 kJ cm^{-2} in 2015 (Figure 4B). Mean UOHC anomaly in July and August 2014 was 16 kJ cm^{-2}, which denotes the beginning of the increasing mean anomaly seen in 2015 and onwards. A decrease in UOHCA was also observed between 2017 and 2018 to approximately 15–23 kJ cm^{-2}, before increasing in mean anomaly in 2019. SST anomalies follow much of the same pattern as UOHCA (Figure 4C). A negative SST anomaly was observed between 2010 and 2013 of −0.5 °C, while SST anomalies above 0.5 °C were observed in July and August between 2014 to 2019. We note that the negative SSTA period coincides with weaker UOHCA than the surrounding years. SST anomalies above 1 °C were observed in 2015 and August 2019. In July and August 2014, the mean SST anomaly in this latitude band was 0.60 °C and 0.61 °C respectively. July and August 2009–2019 SSTA and UOHCA patterns closely followed the observed number of TCs during this time. A negative SSTA and UOHCA anomaly below 10 kJ cm^{-2} was seen when TC counts were low between 2010–2013. An increase in TC counts, with two Hawaiian-landfalling TCs, occurred with a strong positive SSTA of greater than 0.5 °C and UOHCA above 15 kJ cm^{-2}. This observation aligns well with the understanding that wintertime positive El Niño events increase the subsurface heat in the NPO, which can intensify NPO TC activity in the following seasons [33]. Based on the Climate Prediction Center Oceanic Nino Index (ONI), the Julys and Augusts of 2009, 2015, and 2019 were above +0.5 °C (associated with positive El Niño phase), and 2010, 2011, and 2016 were below

−0.5 °C (associated with negative La Niña phase). While the active 2014 NPO TC season, and the contemporaneous SST and UOHC anomalies, are considered to be during 'neutral' conditions not tied to a positive El Niño event, the high TC frequency and the spatial distribution of developing oceanic anomalies leading up to the 2015 El Niño event may prove to be a useful example in furthering our understanding of the relationship between TCs and warming oceanic conditions.

3.2. Storm-Specific Oceanic Conditions

3.2.1. Hurricane Genevieve

Hurricane Genevieve was the first in a sequence of four TCs that formed in the NPO over an 11-day period in late July and early August [34]. Genevieve spent most of its lifecycle within the Eastern and Central NPO oscillating between tropical storm and tropical depression status (Figure 5). After tracking west-southwest of the Hawaiian Islands, around 175° W Genevieve underwent rapid intensification and attained hurricane status on 6 August and major hurricane status on 7 August. While still in the Central NPO, Genevieve was a major hurricane for 6 h with 67 ms^{-1} wind speeds minutes before crossing the International Date Line (180°) into the Western NPO basin. Records from the Japanese Meteorological Agency indicated that Genevieve reached 72 ms^{-1} winds quickly [34] within the Western NPO, and was classified as a super-typhoon, the equivalent of a Category 5 hurricane. An extremely rare occurrence, Genevieve existed in all three Pacific Ocean hurricane basins at some instance throughout its entire lifecycle. Based off of the NE/NC HURDAT2 record, this has only occurred seven other times since 1949.

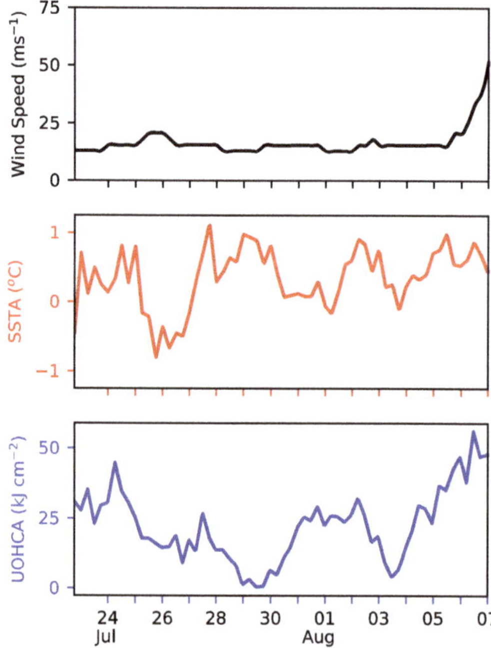

Figure 5. Along-track time series record for Hurricane Genevieve through (**Top**) maximum sustained wind speed with TC intensity category indicated, (**Center**) SST anomaly, and (**Bottom**) UOHC anomaly.

The along-track analysis for Genevieve indicates that despite an extended period of weak TC maintenance, SSTs along its path remained above a positive SST anomaly for most of its duration as a TC in the East and Central NPO basins (Figure 5). Interestingly, SST anomaly at the time of genesis was found to be slightly negative, at −0.45 °C. SST anomaly along Genevieve's path was largely positive, between 0° and 1 °C, with brief periods of negative SST anomaly, with the largest negative anomaly lasting from 05:00 Coordinated Universal Time (UTC) 25 July to 01:00 UTC 27 July, but never lower than −0.8 °C. Coincidentally, this period also corresponded to Genevieve's brief interlude spent at tropical storm status prior to rapid intensification. Starting at 00:00 UTC on 6 August, Genevieve transitioned from a weak tropical storm towards major hurricane status (50 ms^{-1} and greater). During this time, SST anomaly followed much of the same pattern as previously observed (Figure 5), ranging from a positive 0.5 °C to 0.8 °C SST anomaly, which would likely not account for the rapid change of TC intensity. In this regard, we consider the UOHC anomaly and note that UOHC anomaly along-track remained positive throughout. At genesis, UOHC anomaly was 30.9 kJ cm^{-2}. Throughout Genevieve's duration, UOHC anomaly oscillated between 0 and 35 kJ cm^{-2} until 3 August, when Genevieve subsequently tracked over a region of higher positive UOHC anomaly that steadily increased along-track UOHCA to 56 kJ cm^{-2} near 155° W and westward. Looking back to Figures 2 and 3, we observed that the southwestward turn in Genevieve's track brought the TC to the very northern edge of the West Pacific Warm Pool extension of very high UOHC, but also large positive UOHC anomalies at that time as well. We note that this 72-h period of increasing UOHCA prior to rapid intensification on 6 August may have been an oceanic influence on the development of Genevieve into a major hurricane. Storm conditions, SST, and UOHC for Genevieve's duration in the Western Pacific were not investigated in this study.

3.2.2. Hurricane Iselle

Along-track conditions revealed that Iselle formed as a tropical low and traveled over a consistently high positive SST anomaly of 0.5 to 1.0 °C for 36 h after storm genesis on 30 July (Figure 6). Subsequently, Iselle moved into a region of relatively lower positive SST anomaly, hovering around the 0.0 °C anomaly threshold, indicating normal SST conditions in that region of the NPO at this time. Iselle tracked above two regions on 3 August and 5 August of moderately low negative SST anomaly of up to −1 °C, before transitioning back into an area of slightly higher than average SST conditions (Figure 3). We observed that the SST anomaly had more variation in values during this time, ranging from slightly positive anomaly to a moderate negative anomaly. In regards to the UOHC anomaly, genesis occurred in very high UOHCA conditions of 40 kJ cm^{-2}, the highest value observed throughout Iselle's lifespan (Figure 6). UOHC anomaly then dropped to approximately 15 kJ cm^{-2}, shortly after genesis but then increased steadily to 30 kJ cm^{-2} by 1 August. A similar rise and fall of the UOHC anomaly pattern continued over the next few days, with the overall pattern indicating a gradual decline of UOHC anomaly before values dipped into a minor negative UOHC anomaly late on 3 August, not exceeding −3 kJ cm^{-2}. At the time of peak TC wind speed intensity as a Category 4 hurricane, UOHC anomaly ranged from 13 to 14 kJ cm^{-2}. We note that the presence of a positive, but generally declining UOHC anomaly from time of genesis to peak intensification was present along-track for Iselle. Interestingly, UOHC anomaly appeared to increase slightly shortly before landfall on the Puna Coast of the Big Island of Hawaii on 8 August 12:30 UTC.

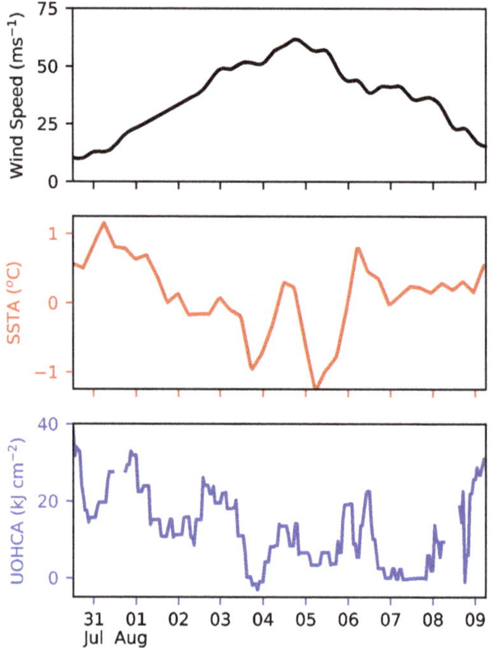

Figure 6. Along-track time series record for Hurricane Iselle through (**Top**) maximum sustained wind speed with TC intensity category indicated, (**Center**) SST anomaly, and (**Bottom**) UOHC anomaly. Landfall is denoted by a break in the UOHC anomaly record. Gap in UOHC anomaly coverage occurred during landfall event.

3.2.3. Hurricane Julio

Time series analysis indicated that Julio formed in a region of slight positive SST anomaly of 0.5 °C (Figure 7), which steadily increased to 1.0 °C before it encountered an oscillating pattern of weak positive to weak negative SST anomalies along-track throughout peak intensity. Julio experienced a weak second increase in TC intensity briefly on 13–14 August. At this time the SST anomaly experienced wide swings between negative anomalies of −0.8 °C and positive anomalies of 1.1 °C. The along-track SST anomaly after this second peak intensity phase then increased to a peak of 2.0 °C, the largest SST anomaly seen along-track for any of the three storms of interest. Figure 3B reveals that this SST anomaly resulted from the large swath of very high SST anomaly throughout the extratropical band north of the Hawaiian Islands in the NPO. Unlike the previous two TCs, Julio's genesis occurred in an area of negative UOHC anomaly of −0.65 kJ cm^{-2} before it quickly increased along-track to a peak of 41 kJ cm^{-2} (Figure 7). UOHC anomaly throughout the first intensification phase for Julio remained positive, averaging between 10 and 20 kJ cm^{-2}. Like Iselle, Julio's peak UOHC anomaly appeared several days prior to peak intensification, with a varying but declining UOHC anomaly afterwards. After Julio reached peak TC intensity as a Category 3 hurricane, along-track UOHC anomaly dropped to near 0 kJ cm^{-2} for approximately 72 h, before entering a region of comparatively low positive UOHC anomaly between 10 to 15 kJ cm^{-2} at the same time as Julio's second increase in intensity to a Category 1 hurricane on 14 August. UOHC anomaly then decreased quickly on 16 August to 0 kJ cm^{-2} for the remainder of storm duration as Julio dissipated. As with Genevieve and Iselle, we noted the importance of positive UOHC anomaly during the intensification phase of Julio prior to reaching peak intensity.

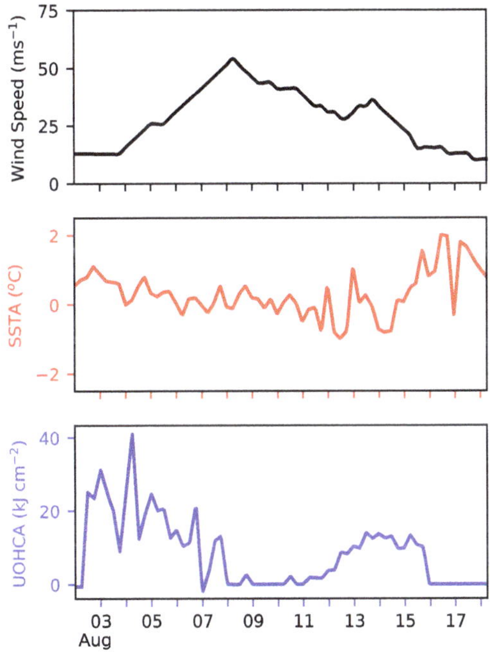

Figure 7. Along-track time series record for Hurricane Julio through (**Top**) maximum stustained wind speed with TC intensity category indicated, (**Center**) SST anomaly, and (**Bottom**) UOHC anomaly. Note: change in scale for SST anomaly (SSTA).

4. Discussion and Conclusions

This research employed SST, SSTA, UOHC, and UOHCA to investigate the anomalous oceanic environment during the generation and mature phases of three major East and Central North Pacific hurricanes in July and August 2014. Basin-wide maps revealed that SST and UOHC were higher than normal over large areas of the Eastern and Central NPO from 0–40° N during the 2014 hurricane season. Maximum SST anomalies occurred along the Mexican coastline and extended offshore to 140° W. The UOHC anomalies in the Eastern NPO did not extend as far north as the SST anomalies but were more spatially concentrated between 5–25° N in the normal generation area for TCs [31]. These anomalous SST and UOHC conditions in the cyclogenesis region between 10–20° N across the NPO may have expanded the area of TCs and hurricane activity to the west and northwest. Hurricane Genevieve formed in the Eastern NPO at 121° W, before crossing the entire breadth of the Central NPO, and reached peak intensity in the Western NPO. Both Hurricanes Iselle and Julio formed in the Eastern NPO before tracking on northwestward paths towards the Hawaiian Islands, including a landfalling strike and an unusual northern trajectory. This was especially true for the region surrounding the Hawaiian Islands and the Central NPO, where maximum positive SST and UOHC anomalies were present north of 30° N into the extratropics. We also make note of the "Blob" present throughout the 2014 hurricane season, where its southern extension is present in the SST/SSTA analyses as a strong +2 °C anomaly.

A few critical observations were made from the along-track analyses. An increase of 50 kJ cm^{-2} in UOHCA was observed prior to and throughout the intensification phase of Genevieve, before tracking into the Western NPO, where further intensification led to the equivalent of a Category 5 hurricane. UOHCA remained high from genesis to peak intensity for both Iselle and Julio, with values typically between 20–30 kJ cm^{-2}. Julio's second brief intensification phase was precluded by an increase in along-track UOHCA from 0 kJ cm^{-2} up to 13 kJ cm^{-2}. The main inference drawn from each of these case studies was the importance of a positive UOHC anomaly present within the development or intensification phase of each of these storms at least 24 h prior to and up to peak intensity, and in Julio's case, a second brief intensification phase. In both Iselle and Julio's intensification phases from cyclogenesis to peak intensity, we noted that the highest UOHCA was observed some time prior to attaining their respective maximum intensities, with positive but declining UOHC anomalies observed afterwards. There may be a variety of factors leading to and influencing this behavior. The oceanic conditions that we have presented are only one component out of several, including the influence of atmospheric circulation, that are involved in the maintenance and peak intensification of any particular TC. More comprehensive research into the importance of positive SST and UOHC anomalies present throughout genesis, intensification, and peak intensity is needed.

A diagnosis of basin-wide SST and UOHC conditions, and their respective anomalies, pointed towards the importance of the location of the 26 °C isotherm. For Hurricane Iselle, this may have been a key component of its trajectory parallel to the Hawaiian Islands, and subsequent landfall. In the case of Hurricane Julio, it is apparent that the storm tracked closely behind Iselle along the monthly mean 26 °C isotherm which was positioned to the northeast and north of Hawaii. A close inspection of oceanic conditions indicates that hurricanes can be sustained in the Eastern NPO with less than optimal oceanic conditions (i.e., when SST and/or UOHC are near the recognized minimum threshold of 26 °C). Although generation of these three hurricanes were observed to occur when SST, SSTA, UOHC, and UOHCA were higher than normal and well above established minimum thresholds, the maintenance of TCs does not appear to require the same level of surface layer temperatures as is characteristic in the Western NPO [16].

We also observed the emerging importance of oceanic conditions in the Central NPO through analyses of high SST and UOHC. In July and August 2014, the surface location of the 26 °C isotherm followed 30° N as far east as 155° W and to the north of the Hawaiian Islands. East of 140° W, we observed the 26 °C isotherm at the surface retreat to approximately 18–20° N as the influence of the much cooler California current and cool surface upwelling along the western U.S. coastline exerted dominant control over surface conditions. Positive SST anomalies were extant throughout the Central NPO, especially the southern and southwestern edge of the "Blob" around 30° N between 180° and 135° W. Our observed anomalies agree well with anomalies reported by Bond et al. [32] for February 2014, hinting at the long-lasting nature of this anomalous feature. Our analysis of Figures 2 and 3 point towards a strengthening of the southern portion of SSTA in the "Blob" in August of 2014.

The penetration of deep warm water from the West Pacific Warm Pool is also visible in both UOHC and UOHCA, and may have contributed to the final rapid intensification momentum of Hurricane Genevieve into the Western NPO. UOHC in this low tropical region exceeded 200 kJ cm^{-2}, with anomalies ranging from 20 to 80 kJ cm^{-2}, confirming the above-average potential available energy provided by the surface ocean during this hurricane season.

It is also important to discuss the El Niño conditions prior to the 2014 hurricane season. Indications of strong positive UOHC and SST anomalies in early 2014 pointed towards the development of a potent El Niño event, with initial similarities to the 1997–1998 El Niño [35]. However, the anticipated strong El Niño phase failed to initiate until the following year. While the 2014 NPO TC season saw above-average activity, prolonged UOHC and SST anomalies observed throughout the 2014 season persisted into 2015, and saw the genesis of record-breaking Hurricane Patricia in the Eastern NPO [36] during a positive El Niño event. Future research would benefit from additional case study analyses to further understand the linkages between upper ocean and SST anomalies as critical factors in the genesis and intensification of major TCs within the Eastern and Central NPO regions.

Supplementary Materials: The following are available online at http://www.mdpi.com/2077-1312/8/4/288/s1, Figure S1: Along-track SST for Hurricanes Genevieve, Iselle, and Julio at 0 to 5 days prior data extraction; Figure S2: Along-track depth of the 26 °C isotherm for Hurricanes Genevieve, Iselle, and Julio.

Author Contributions: Conceptualization, V.L.F. and N.D.W.; methodology, V.L.F. and N.D.W.; formal analysis, V.L.F. and N.D.W.; investigation, V.L.F., N.D.W., and I.-F.P.; data curation, I.-F.P.; writing—original draft preparation, V.L.F. and N.D.W.; writing—review and editing, V.L.F., N.D.W., and. I.-F.P.; visualization, V.L.F. All authors have read and agreed to the published version of the manuscript.

Funding: V.L.F. was funded by the Department of Oceanography and Coastal Sciences and the Earth Scan Laboratory, Louisiana State University, Baton Rouge, Louisiana. I.-F. P. is supported by Taiwan's Ministry of Science and Technology Grant MOST 107-2111-M-008-001-MY3.

Acknowledgments: Portions of this work were presented in thesis form in the fulfillment of the MS requirements for V.L.F. from Louisiana State University. V.L.F. thanks A. Haag of the Earth Scan Laboratory at Louisiana State University for their insight and programming assistance throughout this research study.

Conflicts of Interest: The authors declare no conflict of interest.

References

1. Rappaport, E. Fatalities in the United States from Atlantic Tropical Cyclones: New Data and Interpretation. *Bull. Amer. Meteor. Soc.* **2014**, *95*, 341–346. [CrossRef]
2. Berg, G.D. Eastern Pacific Hurricane Season. Technical Report; National Hurricane Center Annual Summary, 2015. Available online: https://www.nhc.noaa.gov/data/tcr/summary_epac_2014.pdf (accessed on 29 August 2015).
3. Bell, G.D.; Halpert, M.S.; Schnell, R.C.; Higgins, R.W.; Lawrimore, J.; Kousky, V.E.; Tinker, R.; Thiaw, W.; Chelliah, M.; Artusa, A. Climate assessment for 1999. *Bull. Amer. Meteor. Soc.* **2000**, *81*, S1–S50. [CrossRef]
4. Herring, S.C.; Hoerling, M.P.; Kossin, J.P.; Peterson, T.C.; Stott, P.A. Explaining Extreme Events of 2014 from a Climate Perspective. *Bull. Amer. Meteor. Soc.* **2015**, *96*, S1–S172. [CrossRef]
5. Businger, S.; Nogelmeier, M.P.; Chinn, P.W.; Schroeder, T. Hurricane with a History: Hawaiian Newspapers Illuminate an 1871 Storm. *Bull. Amer. Meteor. Soc.* **2018**, *99*, 137–147. [CrossRef]
6. Berg, R.; Kimberlain, T. The 2014 Eastern North Pacific Hurricane Season: A Very Active Season Brings Devastation. *Weatherwise* **2015**, *68*, 36–45. [CrossRef]
7. Leipper, D.; Volgenau, D. Hurricane Heat Potential of the Gulf of Mexico. *J. Phys. Oceanogr.* **1972**, *2*, 218–224. [CrossRef]
8. DeMaria, M.; Kaplan, J. Sea surface temperature and the maximum intensity of Atlantic tropical cyclones. *J. Climate* **1994**, *7*, 1325–1334. [CrossRef]
9. Whitney, L.; Hobgood, J. The relationship between sea surface temperature and maximum intensities of tropical cyclones in the eastern North Pacific. *J. Climate* **1997**, *10*, 2921–2930. [CrossRef]
10. Webster, P.; Holland, G.; Curry, J.; Chang, H.R. Changes in Tropical Cyclone Number, Duration, and Intensity in a Warming Environment. *Science* **2005**, *309*, 1844–1846. [CrossRef]
11. Goni, G.; Trinanes, J. Ocean thermal structure monitoring could aid in the intensity forecast of tropical cyclones. *EOS Trans. Am. Geophys. Union* **2003**, *83*, 573–578. [CrossRef]
12. Cione, J. The Relative Roles of the Ocean and Atmosphere as Revealed by Buoy Air-Sea Observations in Hurricanes. *Mon. Weather Rev.* **2015**, *143*, 904–913. [CrossRef]
13. Shay, L.; Goni, G.; Black, P. Effects of a warm oceanic feature on Hurricane Opal. *Mon. Weather Rev.* **2000**, *128*, 1366–1383. [CrossRef]
14. Mainelli, M.; DeMaria, M.; Shay, L.; Goni, G. Application of oceanic heat content estimation to operational forecasting of recent Atlantic category 5 hurricanes. *Weather Forecast.* **2008**, *23*, 3–16. [CrossRef]
15. Shay, L.; Brewster, J. Oceanic Heat Content Variability in the Eastern Pacific Ocean for Hurricane Intensity Forecasting. *Mon. Weather Rev.* **2010**, *138*, 2110–2131. [CrossRef]
16. Lin, I.I.; Pun, I.F.; Wu, C.C. Upper-Ocean Thermal Structure and the Western North Pacific Category 5 Typhoons. part II: Dependence on Translation Speed. *Mon. Weather Rev.* **2009**, *137*, 3744–3757. [CrossRef]
17. Fisher, E. Hurricanes and the Sea-Surface Temperature Field. *J. Meteor.* **1958**, *15*, 328–333. [CrossRef]
18. Price, J. Upper Ocean Response to a Hurricane. *J. Phys. Oceanogr.* **1981**, *11*, 153–174. [CrossRef]
19. Wroe, D.; Barnes, G. Inflow Layer Energetics of Hurricane Bonnie (1998) near Landfall. *Mon. Weather Rev.* **2003**, *131*, 1600–1612. [CrossRef]
20. Walker, N.; Leben, R.; Pilley, C.; Shannon, M.; Herndon, D.; Pun, I.F.; Lin, I.I.; Gentemann, C. Slow translation speed causes rapid collapse of northeast Pacific Hurricane Kenneth over cold core eddy. *Geophys. Res. Lett.* **2014**, *41*, 7595–7601. [CrossRef]

21. Gentemann, C.; Meissner, T.; Wentz, F. Accuracy of Satellite Sea Surface Temperatures at 7 and 11 GHz. *IEEE Trans. Geosci. Remote Sens.* **2010**, *48*, 1009–1018. [CrossRef]

22. Wentz, F.; Gentemann, C.; Smith, C.; Chelton, D. Satellite measurements of sea surface temperature through clouds. *Science* **2000**, *288*, 847–850. [CrossRef]

23. Ford, V.L. The Role of Upper Ocean Heat Content and Sea Surface Temperature on Northeast Pacific Hurricane Evolution During Average and Active Years. Master's Thesis, Louisiana State University, Baton Rouge, LA, USA, 2016.

24. Banzon, V.; Smith, T.; Chin, T.; Liu, C.; Hankins, W. A long-term record of blended satellite and In Situ sea-surface temperature for climate monitoring, modeling and environmental studies. *Earth Syst. Sci. Data* **2016**, *8*, 165–176. [CrossRef]

25. Xue, Y.; Smith, T.; Reynolds, R. Interdecadal Changes of 30-Yr SST Normals during 1871-2000. *J. Clim.* **2003**, *16*, 1601–1612. [CrossRef]

26. Pun, I.F.; Lin, I.I.; Ko, D. New generation of satellite-derived ocean thermal structure for the western North Pacific typhoon intensity forecasting. *Oceanography* **2014**, *121*, 109–124. [CrossRef]

27. Goni, G.; Kamholz, J.; Garzoli, S.; Olson, D. Dynamics of the Brazil-Malvinas Confluence based on inverted echo sounders and altimetry. *JGR Oceans* **1996**, *101*, 16273–16289. [CrossRef]

28. Pun, I.F.; Price, J.; Jayne, S. Satellite-Derived Ocean Thermal Structure for the North Atlantic Hurricane Season. *Mon. Weather Rev.* **2016**, *144*, 877–896. [CrossRef]

29. Landsea, C.; Franklin, J. Atlantic Hurricane Database Uncertainty and Presentation of a New Database Format. *Mon. Weather Rev.* **2013**, *141*, 3576–3592. [CrossRef]

30. Elsner, J.; Jagger, T. *Hurricane Climatology: A Modern Statistical Guide Using R*, 1st ed.; Oxford University Press: New York, NY, USA, 2013.

31. Farfan, L. Eastern Pacific Tropical Cyclones and Their Impact Over Western Mexico. In *Experimental and Theoretical Advances in Fluid Dynamics*; Klapp, J., Cros, A., Velasco Fuentes, O., Stern, C., Rodriguez Meza, M., Eds.; Springer: Berlin/Heidelberg, Germany, 2012; pp. 135–148.

32. Bond, N.; Cronin, M.; Freeland, H.; Mantua, N. Causes and impacts of the 2014 warm anomaly in the NE Pacific. *Geophys. Res. Lett.* **2015**, *42*, 3414–3420. [CrossRef]

33. Jin, F.F.; Boucheral, J.; Lin, I.I. Eastern Pacific tropical cyclones intensified by El Niño delivery of subsurface ocean heat. *Nature* **2014**, *516*, 82–85. [CrossRef]

34. Beven, J.; Birchard, T. Hurricane Genevieve (EP072014). Available online: https://www.nhc.noaa.gov/data/tcr/EP072014_Genevieve.pdf (accessed on 15 May 2016).

35. McPhaden, M. Playing hide and seek with El Niño. *Nat. Clim. Chang.* **2015**, *5*, 791–795. [CrossRef]

36. Foltz, G.; Balaguru, K. Prolonged El Niño conditions in 2014–2015 and the rapid intensification of Hurricane Patricia in the eastern Pacific. *Geophys. Res. Lett.* **2016**, *43*, 10347–10355. [CrossRef]

Journal of
Marine Science and Engineering

Article

The North Equatorial Countercurrent East of the Dateline, Its Variations and Its Relationship to the El Niño Event

Yusuf Jati Wijaya [1,2], Ulung Jantama Wisha [1,3] and Yukiharu Hisaki [1,*]

1   Department of Physiscs and Earth Sciences, University of the Ryukyus, Senbaru 1, Nishihara-cho, Okinawa 903-0213, Japan; yusufjatiwijaya@gmail.com (Y.J.W.); ulungjantama@kkp.go.id (U.J.W.)
2   Department of Oceanography, Faculty of Fisheries and Marine Science, Diponegoro University, Semarang 50275, Jawa Tengah, Indonesia
3   Research Institute for Coastal Resources and Vulnerability, Ministry of Marine Affairs and Fisheries, Jl. Raya Padang-Painan KM. 16, Bungus, Padang 25245, Sumatera Barat, Indonesia
*   Correspondence: hisaki@sci.u-ryukyu.ac.jp; Tel.: +81-98-895-8515

Abstract: Using forty years (1978–2017) of Ocean Reanalysis System 4 (ORAS4) dataset, the purpose of this study is to investigate the fluctuation of the North Equatorial Countercurrent (NECC) to the east of the dateline in relation to the presence of three kinds of El Niño events. From spring (MAM) through summer (JJA), we found that the NECC was stronger during the Eastern Pacific El Niño (EP El Niño) and the MIX El Niño than during the Central Pacific El Niño (CP El Niño). When it comes to winter (DJF), on the other hand, the NECC was stronger during the CP and MIX El Niño and weaker during the EP El Niño. This NECC variability was affected by the fluctuations of thermocline depth near the equatorial Pacific. Moreover, we also found that the seasonal southward shift of the NECC occurred between winter and spring, but the shift was absent during the CP and MIX El Niño events. This meridional shift was strongly affected by the local wind stress.

Keywords: El Niño; the NECC; east of the dateline; Pacific Ocean

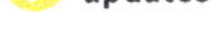
check for
updates

Citation: Wijaya, Y.J.; Wisha, U.J.; Hisaki, Y. The North Equatorial Countercurrent East of the Dateline, Its Variations and Its Relationship to the El Niño Event. *J. Mar. Sci. Eng.* **2021**, *9*, 1041. https://doi.org/10.3390/jmse9101041

Academic Editor: Francisco Pastor

Received: 31 August 2021
Accepted: 20 September 2021
Published: 22 September 2021

Publisher's Note: MDPI stays neutral with regard to jurisdictional claims in published maps and institutional affiliations.

1. Introduction

Situated in the low-latitude North Pacific Gyre, the North Equatorial Countercurrent (NECC) is an eastward-flowing surface current. Its position is centered at the latitude of ~5° N to the west of the dateline [1,2] and further eastward, shifting toward ~7° N in the eastern part of the Pacific [3,4]. The NECC plays a pivotal role in the global atmospheric circulation since it carries on average 20 to 30 Sv (1 Sv is equal to 10^6 m^3 s^{-1}) of surface water of warm pool from the western to the eastern Pacific Ocean [3,5,6]. The NECC is closely related to the Intertropical Convergence Zone, which is crucial to the distribution of nutrients in the tropical Pacific Ocean [7].

The dynamics of the NECC are strictly tied to the El Niño-Southern Oscillation (ENSO) event according to the work of previous studies [1,2,8–12]. This is not surprising, as the NECC is located at the place where the ENSO event is evoked. Satellite and ocean simulation observations have suggested a strong magnitude and a southerly path of the NECC west of the dateline during El Niño, whereas its position shifts to the north and is weakened in La Niña [2,9,11]. The westerly wind associated with Rossby wave variations is highly responsible for those occurrences. East of the dateline, El Niño's effect on the NECC is determined by the type of El Niño itself. Previous observations have separated El Niño events into two different classifications depending on the anomalous pattern of the sea surface temperature (SST) throughout the Pacific Ocean, namely the central Pacific (CP) El Niño and the eastern Pacific (EP) El Niño [13–17]. Some researchers have proposed that EP El Niño's influence on the NECC was greater than that of the CP El Niño east of the dateline [8,10,12]. During the EP El Niño, the NECC tends to strengthen and migrate southward from the phase of development to the mature phase because of the shifting of the curl of wind stress distribution, which leads to changes in thermocline variability.

83

To date, the NECC variations east of the dateline are still an interesting topic to observe, notably their varied reactions to the EP and CP types of El Niño events throughout the phases of developing and maturing. Thus, in this paper, we propose an investigation of the NECC variations to the east of the dateline using the Empirical Orthogonal Function (EOF) approach as the analysis technique to provide a new suggestion or enhance the ideas of previous research. We used the zonal component of the surface current from reanalysis data for 40 years of observations to investigate their spatial and temporal variations, especially during summer and winter, which are often referred to as the stages of developing and maturing from El Niño, respectively [18,19]. This paper is arranged in the following way: Section 2 explained the materials and the methods used; Section 3 contained a summary of the experimental outcomes; finally, in Sections 4 and 5, we conclude the paper with a discussion and conclusion, respectively.

2. Materials and Methods

The Ocean Reanalysis System 4 (ORAS4) dataset retrieved from European Centre for Medium-Range Weather Forecasts (ECMWF) was used in this work, which spans 40 years (1978–2017) and has a 1-degree grid resolution. These data are monthly averages of the zonal and meridional components of the ocean surface current, extending from 4.5° N to 11.5° N and 178.5° W to 70.5° W. The ORAS4 dataset is freely available from 1958 to 2017 and covers the entirety of the Earth. By replacing its predecessor, ORAS3, various recent features and enhancements have been implemented in ORAS4 with respect to the former product, such as model bias correction. For more thorough information on the data features and specifications, see Balmaseda et al. [20] and Balmaseda et al. [21]. Moreover, to explore the explanation of the NECC variations east of the dateline, we utilized the Ocean Reanalysis System 5's (ORAS5's) 20 °C isotherm depth from ECMWF. The 20 °C isotherm depth's spatial and temporal resolutions are 1° and monthly, respectively [22].

We also used the ECMWF Reanalysis v5 (ERA5) surface wind dataset over the Pacific Ocean, provided by the Copernicus Climate Change Service (C3S), to analyze the wind trend during the EP and CP of El Niño. By replacing ERA-Interim reanalysis, the data quality of the ERA5 reanalysis improved. The latitude and longitude grid resolution of these data is a $0.25° \times 0.25°$ grid. The outstanding details of the ERA5 data information and specifications were provided by Hersbach et al. [23] and Molina et al. [24]. The wind data were converted to zonal (τ_x) and meridional (τ_y) wind stress components, obtained using Equations (1) and (2), taken from Kok et al. [25]:

$$\tau_x = \rho_a C_d (u^2 + v^2)^{1/2} u \tag{1}$$

$$\tau_y = \rho_a C_d (u^2 + v^2)^{1/2} v \tag{2}$$

where ρ_a is the air density (1.2 kg/m^3); the drag coefficient, C_d, equaled 1.3×10^{-3}, while u indicates the zonal component of wind, and v represents the meridional component. Additionally, Sverdrup Balance was also estimated to investigate the effects of local wind stress in NECC variability. We determined the Sverdrup zonal transport using the following formula [26]:

$$U = -\frac{1}{\beta} \int \left(\frac{\partial \tau_y}{\partial x} - \frac{\partial \tau_x}{\partial y} \right) dx \tag{3}$$

The National Oceanic and Atmospheric Administration (NOAA) optimum interpolation sea surface temperature (OISST) v2 products were used in this work to compare the horizontal distribution of the SST anomaly in the Pacific Ocean during the different types of El Niño [27]. This data analysis employed a monthly and 1° spatial resolution, which was constructed by combining numerous observations on a regular grid, such as floating devices, satellites, and ship surveys. The Oceanic Niño Index (ONI), expressed by the running 3-month mean SST anomaly in the Niño 3.4 region (which is approximately 5° N to 5° S, 170° W to 120° W), was the significant indicator used to observe the variability of El

Niño and La Niña events. The National Oceanic and Atmospheric Administration (NOAA) supplied this index, which was operated to determine the intensity of the El Niño event. We only observed moderate to very strong events, which were classified as follows: 1.0 to 1.4 for a moderate event, 1.5 to 1.9 for a strong event, and $\geq$2.0 for a very strong event.

Employing wide-area and long-time-series data for observation, the empirical orthogonal function (EOF) appeared to be a tool of convenience for performing the analysis and has been commonly used to extract the dominant variance [28,29]. The EOF analysis produced a pair of dominant spatial patterns along with the corresponding principal component (PC) that displayed the temporal pattern. To begin, we eliminated the annual cycle from the data and then computed the first three EOFs for JJA, SON, DJF, and MAM of the zonal component of the surface current to identify the dominant modes of the NECC variability in the summer, fall, winter, and spring, respectively, over 40 years.

3. Results

We averaged the zonal velocity component of the ORAS4 to the east of the dateline in the Pacific Ocean, which we display in Figure 1a. The eastward-flowing NECC was denoted by the positive value of zonal velocity, which laid between the latitudes of 4° N and 10° N. This region represented the mean pathway of the NECC, which looked uniform to the east of the dateline. Several earlier investigations reported that the NECC pathway was broader to the west of the dateline and narrower to the east of the dateline [8,9]. Moreover, the NECC's average eastward velocity attained its peak between 160° W and 130° W.

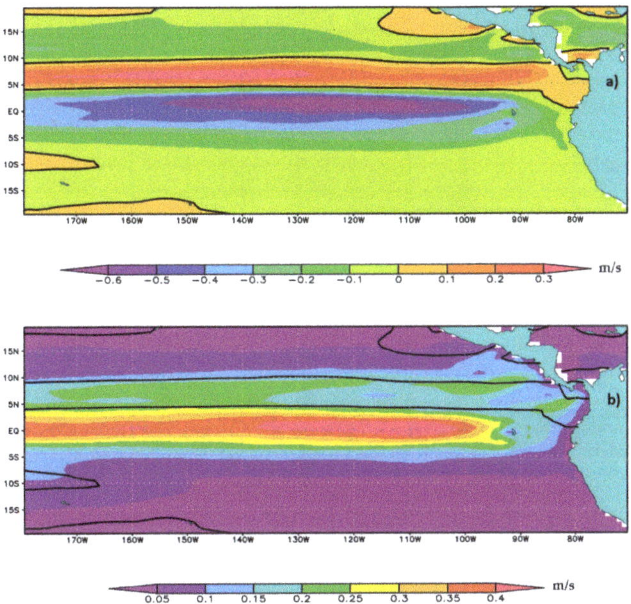

Figure 1. Mean map of (**a**) zonal velocity of ORAS4 surface current and (**b**) the corresponding standard deviation, calculated over the period of 1978 to 2017.

The standard deviation of the zonal velocity from the ECMWF ORAS4 data is presented in Figure 1b, which was computed from January 1978 to December 2017. A large standard deviation was observed just south of the NECC, widely known as south equatorial current (SEC); the standard deviation over the 40-year period was 0.4 m s^{-1}. At the same time, the standard deviation in the NECC region was lower (less than 0.25 m s^{-1}). This result suggested that in order to reduce the interference of the SEC signal, the EOF analysis in the next figure would cover the region between the latitude of 3.5° N to 11.5° N.

3.1. EOF Analysis

The EOF spatial patterns of the NECC for the boreal summer (only the results of the first two EOF modes) are shown in Figure 2. The accumulative explained variance was 62.4% by the first two modes. The first mode (EOF1) explained 41.6% of the total variance. This mode displayed a positive signal region throughout 3.5° N to 8° N that was centered in the central Pacific. In the eastern Pacific, this signal gradually shifted northward to 9° N. The related time series (PC1) showed a strong association between the NECC and ENSO events. Almost every El Niño or La Niña event caused the NECC to strengthen or weaken, as evidenced by the positive or negative phase, respectively, in Figure 2B. A slight difference was found by comparing the EP and CP types of El Niño. For the period of 1978 to 2017, five events of El Niño of the EP type were identified and are shown in Table 1. By contrast, four events were recorded in the case of the CP type [30,31]. From the PC1, the EP El Niño frequently produced a more major positive phase than the CP-El Niño (except for the 1986/1987 event), implying that the NECC was more powerful during the EP type. Furthermore, the second mode (EOF2) explained 20.8% of the total observations. Spatially, the positive signal was found further southward than that of the first mode (EOF1), with the variability maxima located in the area between 150° W and 130° W. The corresponding PC2 showed less of a relationship to the ENSO event.

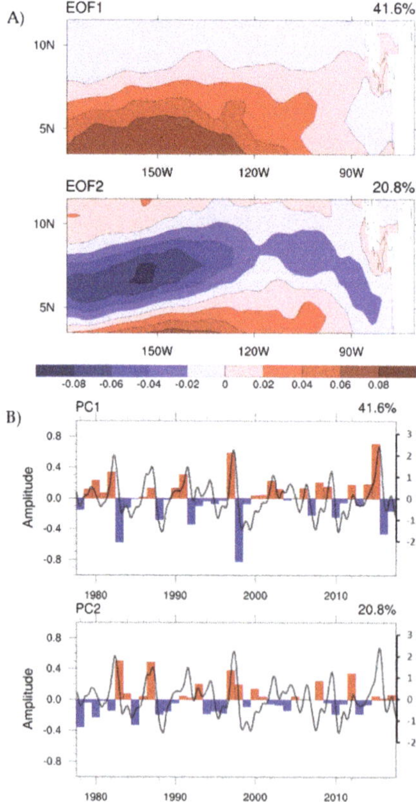

Figure 2. Two leading EOF analyses of the zonal component of surface current for summer (JJA): (**A**) spatial pattern and (**B**) time series of the principal component superimposed with 6 months low-pass filtered NIÑO 3.4 index. Red and blue bars represented the NECC's strengthening and weakening, respectively.

Table 1. El Niño events ranging from moderate to very strong from 1978 to 2017 and their classification based on the Niño methods.

Type	Moderate	Strong	Very Strong
CP El Niño	1994/1995, 2002/2003, 2009/2010	1987/1988	-
EP El Niño	1986/1987	1991/1992	1982/1983, 1997/1998, 2015/2016

The first two leading EOF modes with their corresponding time series (PC) for fall (SON) are plotted in Figure 3, which explained 63.7% and 9.2% of the observed variability, respectively. The first EOF pattern (EOF1) showed a similar pattern to that of the EOF1 for summer (Figure 2A), and the positive signal was found near the equator. From the corresponding PC1, we can see roughly that the variability was very much related to the ENSO event. Unlike the EOF1 for summer, the NECC's reaction to the EP and CP types of El Niño looked no different in this season. Every El Niño event produced a positive phase, while the La Niña produced a negative phase. Moreover, the second EOF2 pattern showed a negative signal that seemed to merge with the positive signal in the north. The associated PC time series (PC2) exhibited that the variability after the 2000s produced the negative phase more.

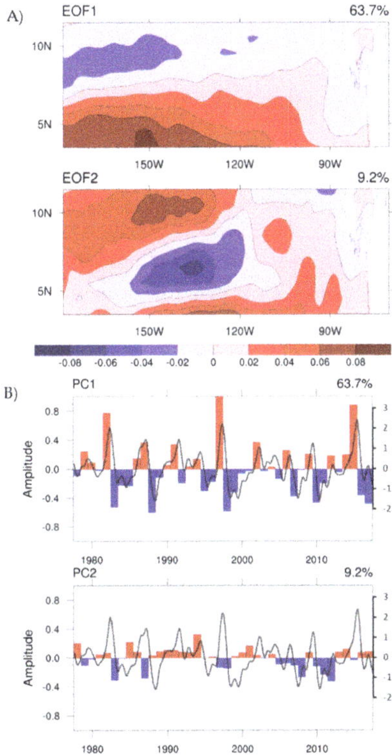

Figure 3. Same as Figure 2, but for fall (SON): (**A**) spatial pattern and (**B**) time series of the principal component superimposed with 6 months low-pass filtered NIÑO 3.4 index. Red and blue bars represented the NECC's strengthening and weakening, respectively.

Figure 4 shows the first two EOF modes of winter (DJF), which explained 30.3% and 11.8% of the total variance of the zonal velocity of the ORAS4 surface current. Spatially, the first leading EOF (EOF1) was maximal near the equator and was centered between the longitudes of 150° W and 120° W. In the eastern Pacific, the positive signal was found around 9° N, similar to the pattern that was shown in the EOF1 for summer. The corresponding PC1 mode indicated a strong relationship to the cold event, portrayed by the negative phase that consistently appeared in each La Niña year. Meanwhile, in relation to the El Niño, the positive phase arose in the years 1987/1988, 1994/1995, 2002/2003, 2009/2010, 1986/1987, 1991/1992, and 2015/2016, of which the last three belonged to the EP type of El Niño and the remainder were CP type. For the winter, these results implied that the CP El Niño generated a relatively stronger NECC than the EP-El Niño. This suggestion is of particular interest since the presence of three EP El Niño events was linked to the intensification of NECC. Nevertheless, Hu et al. [17], Paek et al. [32], and Zhang et al. [33] reported that the El Niño event of 1991/1992 and 2015/2016 were classified as a combination of both CP and EP events, commonly known as the MIX El Niño type, with a wide spread of the SST anomaly to the east of the dateline in the Pacific Ocean. Perhaps for this reason, they appeared as positive phases in the PC1 variability. By contrast, for the 1986/1987 event, Chen and Li [34] reported that the EP El Niño of 1986/1987 was a special EP El Niño that generated strong westerly wind events in the winter and re-evolved into a subsequent El Niño event. By contrast, the second leading mode (EOF2) showed a positive signal further northward. The related time series (PC2) exhibited decreased year-to-year variability beginning in the early 2000s. This result implies that the NECC has undergone weakening in the last two decades. Furthermore, the relationship with the ENSO event was much weaker in PC2.

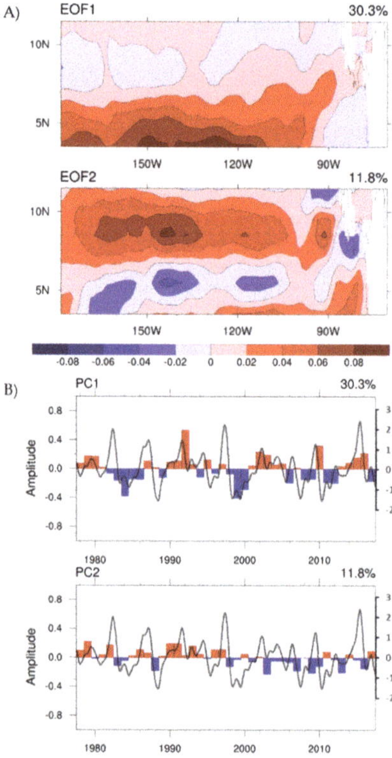

Figure 4. Same as Figure 2, but for winter (DJF): (**A**) spatial pattern and (**B**) time series of the principal component superimposed with 6 months low-pass filtered NIÑO 3.4 index. Red and blue bars represented the NECC's strengthening and weakening, respectively.

As shown in Figure 5, the spring EOF1 pattern (MAM) explained 51.1% of the overall variance. The EOF1 exhibited a dipole spatial pattern, with the positive signal centered at a latitude of ~4.5° N. In the time series, the corresponding PC1 was highly correlated with the EP type of El Niño event. When the EP and MIX types started to develop in the springs of 1982, 1991, 1997, and 2015, they produced an obvious positive phase. The CP type, on the other hand, resulted in a negative phase and a minor positive phase. This result suggests that the EP and MIX El Niño types produced a more powerful NECC than the CP El Niño. On the other hand, in the second EOF (EOF2) mode, the zonal velocities showed a positive-negative-positive pattern with variability maxima north of 3.5° N, which only accounted for 10.5% of the total variance. The related time series PC2 denoted no significant relationship with the ENSO event.

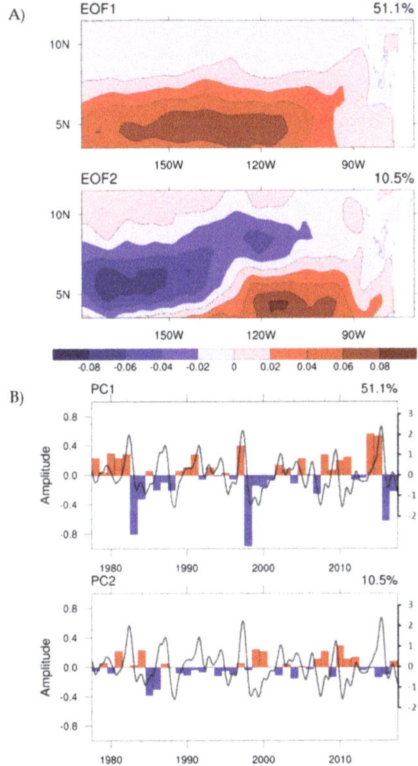

Figure 5. Same as Figure 2, but for spring (MAM): (**A**) spatial pattern and (**B**) time series of the principal component superimposed with 6 months low-pass filtered NIÑO 3.4 index. Red and blue bars represented the NECC's strengthening and weakening, respectively.

3.2. Meridional Shifting of the NECC

Next, to explore the variability of the NECC to the east of the dateline over 40 years, we plotted the zonal component as time against latitude using the meridional line at 150° W as a reference (Figure 6). The positive value was considered the eastward flow, showing variations in the NECC's path. Seasonally, an equatorward shift was observed to occur frequently in mid-winter to spring, but this event was absent in certain years. An El Niño event is capable of inhibiting an equatorward shift of the NECC only if certain criteria are met.

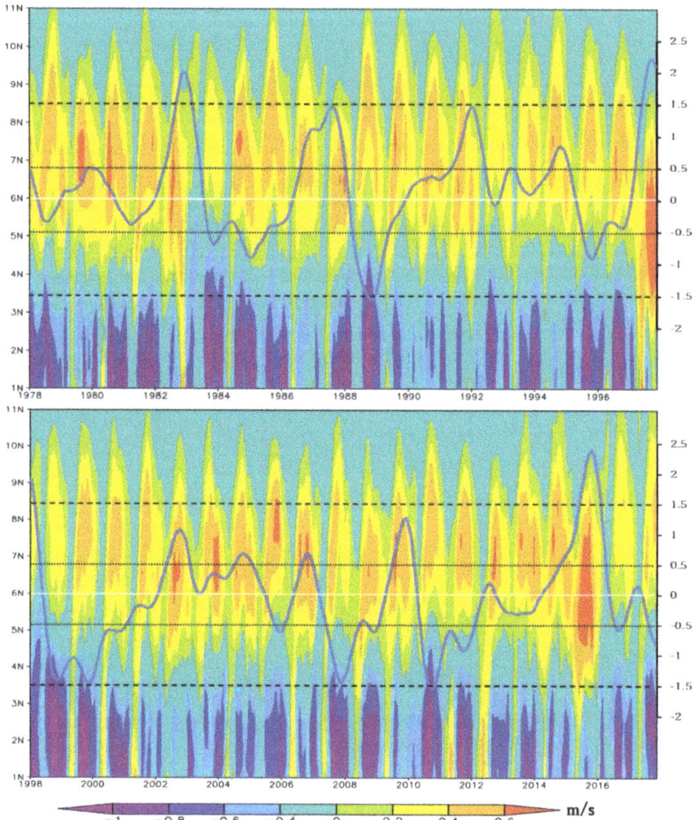

Figure 6. Latitude–time diagram of the zonal component of surface current at the meridional line 150° W, 6-month low-pass filtered NIÑO 3.4 index overlaid (blue line). Upper panel for January 1978 to December 1997 and lower panel for January 1998 to December 2017.

The absence of an equatorward shift corresponded to moderate to very strong El Niño events, but this is not the only requirement to interfere with such a shift. The current article classified El Niño event into three types: EP, CP, and MIX types, as suggested by a number of prior studies [14–17,33]; however, only two forms of El Niño may prevent the NECC so that it does not shift toward the equator, namely the CP and MIX types. Over 40 years, the CP El Niño and MIX El Niño events occurred in the years 1987/1988, 1994/1995, 2002/2003, 2009/2010, 1991/1992, and 2015/2016. The equatorward shift did not take place during the stage of maturing of the events of the CP and MIX types. By contrast, the NECC still shifted toward the equator in the EP type, which was recorded in the year 1982/1983, 1986/1987, and 1997/1998. Furthermore, in the events of 1982/1983 and 1997/1998, which fell into the very strong El Niño category, the equatorward shift occurred earlier in the winter.

The investigation of the wind and SST distribution over the near-equatorial region of the Pacific is critical for a better understanding of the various patterns that formed during the three types of El Niño events. We averaged the components of wind stress and SST over the winter for each kind of El Niño, which are displayed in Figure 7. This figure depicts substantial variation in the distribution of SST and wind stress in the CP, MIX, and EP types of El Niño. All kinds of El Niño were characterized by a positive anomaly of the SST across the equatorial region of the Pacific, and the distinction was in the position of the highest positive anomaly of the SST [14,16,17,33]. The smallest magnitude of the

SST anomaly occurred during the CP type, with the peak located near a longitude of 170° W, accompanied by anomalously westerly wind stress in the southern part of the central equatorial region. As for the MIX El Niño, the highest anomaly of the SST was located around 20° further east than that of the CP type. Strong northerly and westerly wind stress was found in the region between 180° E and 150° W. Moreover, the area of maximum SSTA for the EP El Niño was larger and farther east than those of the CP and MIX types, followed by a significant anomaly of northerly and westerly wind stress in the central-eastern equatorial Pacific. This result also suggests that under all forms of El Niño, the trade wind over the equatorial Pacific was weaker.

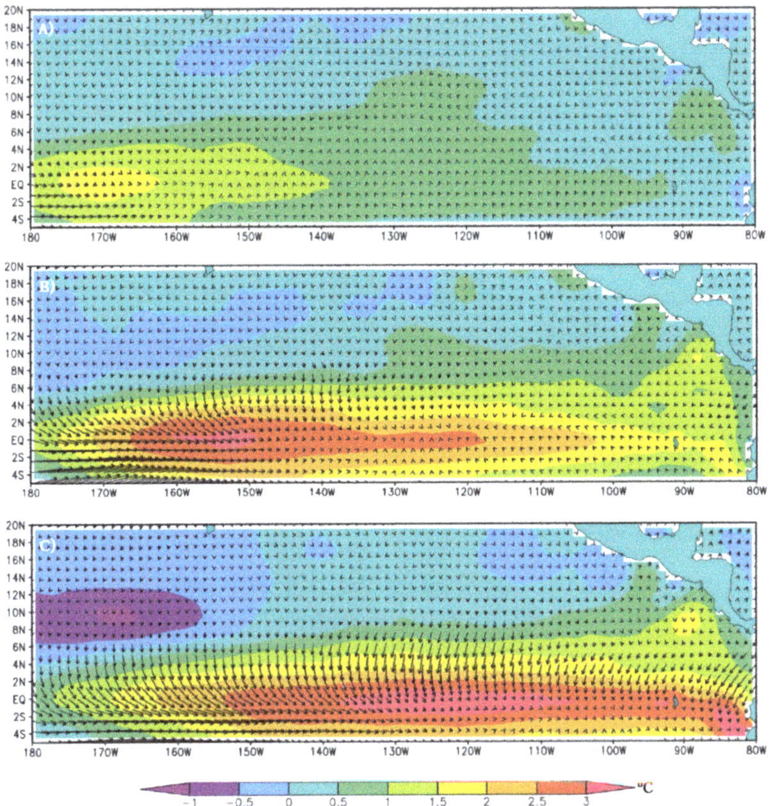

Figure 7. The mean surface wind stress anomaly (vectors) with SST field (shaded) to the east of the dateline during the winter of El Niño events of the (**A**) CP type, (**B**) MIX type, and (**C**) EP type.

Furthermore, to examine the relationship between wind stress distribution and NECC variations, we computed the curl of the wind stress using the zonal and meridional components, then performed Sverdrup balance (SB) analyses. Based on the ECMWF 0.25° monthly wind component of the ERA5, Figure 8 depicts a comparison of the average winter SB in each kind of El Niño event. Significant differences were observed when comparing eastward transport among the CP, MIX, and EP types, which is represented by a positive value of SB. The eastward transport near 0 latitude for the CP El Niño was positioned at ~4° N in the central-eastern Pacific and proceeded northward at a latitude of ~7° N as it approached land. The MIX El Niño's eastward transport was found at 1° N in the central Pacific and progressively migrated northward at a latitude of 3° N in the eastern Pacific, while it laid at ~7° N in the easternmost Pacific Ocean. As for the EP El Niño, this eastward transport was migrated to the south of ~1° S in the central Pacific Ocean and stayed at ~2° N in the eastern Pacific. Moreover, as happened in the CP and MIX types, the eastward-flow transport was similarly positioned around ~7° N latitude in the easternmost Pacific Ocean. Thus, this SB analysis revealed the meridional shift of the eastward-flowing NECC, which indicated good agreement with the analysis of the NECC variations in Figure 6.

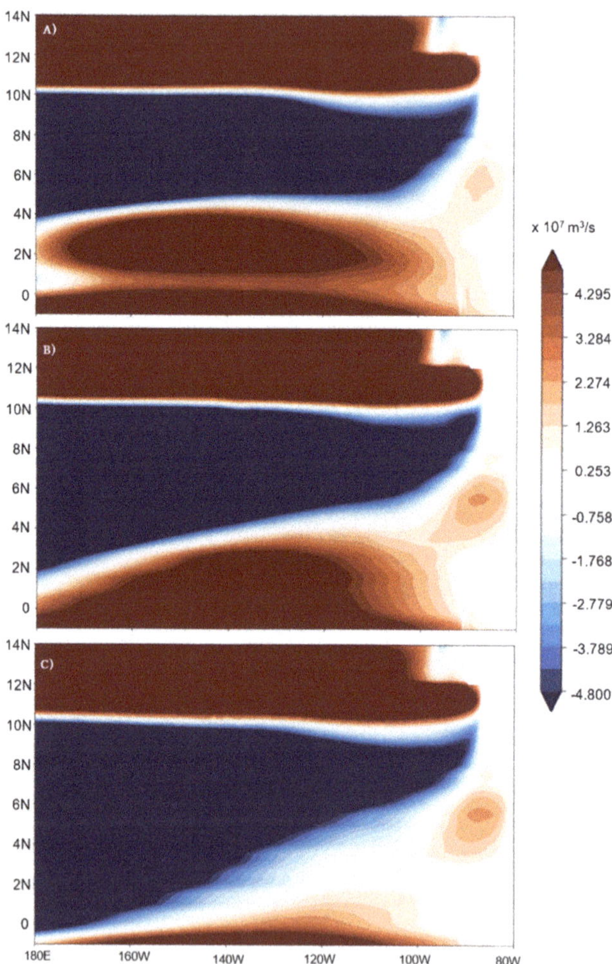

Figure 8. Average of Sverdrup balance (SB) in 10^7 m^3/s during the winter of El Niño events of the (**A**) CP type, (**B**) MIX type, and (**C**) EP type. The colors red and blue represent east and west transport flow, respectively.

4. Discussion

The current paper investigated the NECC variability to the east of the dateline, which was found to be closely associated with the ENSO event. We used the zonal velocity of surface current data derived from the ECMWF from 1978 to 2017 on a regular grid with a $1° \times 1°$ spatial resolution. We applied the EOF analysis to each season, which was broken down into the winter (December to February), spring (March to May), summer (June to August), and fall (September to November). Interestingly, the dominant seasonal pattern of NECC was obtained from this analysis. We also employed the SST data from NOAA and the ERA5 wind data from the ECMWF.

For all seasons, the EOF analysis revealed that the NECC showed a tendency to strengthen in El Niño, while it tended to weaken in La Niña. However, El Niño, categorized into the EP, CP, and MIX types, produced a quite different impact on the NECC each season. For the spring and summer, which are frequently linked with the developing phase, the NECC was substantially more powerful during the EP and MIX types compared with the CP type. By contrast, the EP type weakened the NECC during the winter, which is often considered as the mature stage, whereas it became stronger during the CP and MIX types. Unlike other seasons, the relatively equal impact of the EP and CP types of El Niño on the NECC was observed during the fall season. The NECC indicated a tendency to strengthen in all forms of El Niño. The fluctuation of isotherm depth at 20° (D20) could have caused this NECC variability around the Pacific Ocean (Figures 9 and 10), which was influenced by the wind stress distribution [35]. Through the Ekman transport process, the summer of the EP El Niño raised the thermocline in the northwestern Pacific while lowering it in the eastern Pacific, producing a meridional gradient of sea level, and resulted in a more robust NECC (Figure 9A). During the CP El Niño, on the other hand, the decreased sea level gradient between the northwestern and east Pacific led to a small strengthening of the NECC (Figure 9B). As the CP El Niño entered its mature phase, the shoaling of the thermocline occurred in the northwestern Pacific and then increased the meridional gradient of sea level. As a result, the NECC tended to strengthen (Figure 10B). Moreover, for the EP type, the thermocline shoaling (deepening) took place along the western to the central (eastern) Pacific Ocean, causing the NECC to weaken (Figure 10A). This weaker NECC during the EP El Niño's winter was unexpected and has never been reported in previous studies. Additionally, the thermocline, which occurred in the western Pacific to the east of the dateline (160° W), was raised in January 1983 and 1998 (Figure 11). In consequence, the NECC became weaker. However, its exact mechanism is still undetermined; therefore, we leave for further research the investigation of how the winter of EP El Niño could generate a weaker NECC.

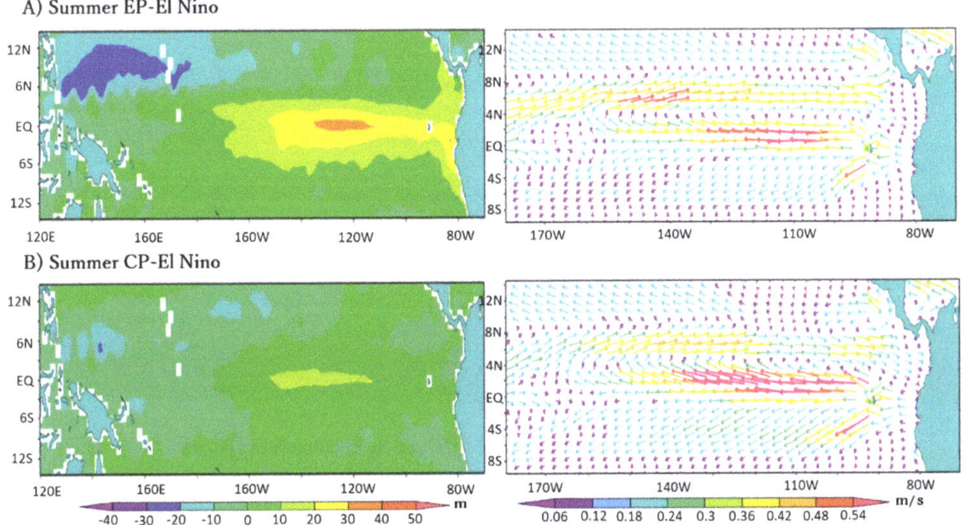

Figure 9. Composite maps of 20 °C isotherm depth (D20) anomalies (left panel) and ocean surface current (right panel) for the summer from (**A**) the EP El Niño and (**B**) CP El Niño. Negative (Positive) D20 anomalies denote shoaling (deepening).

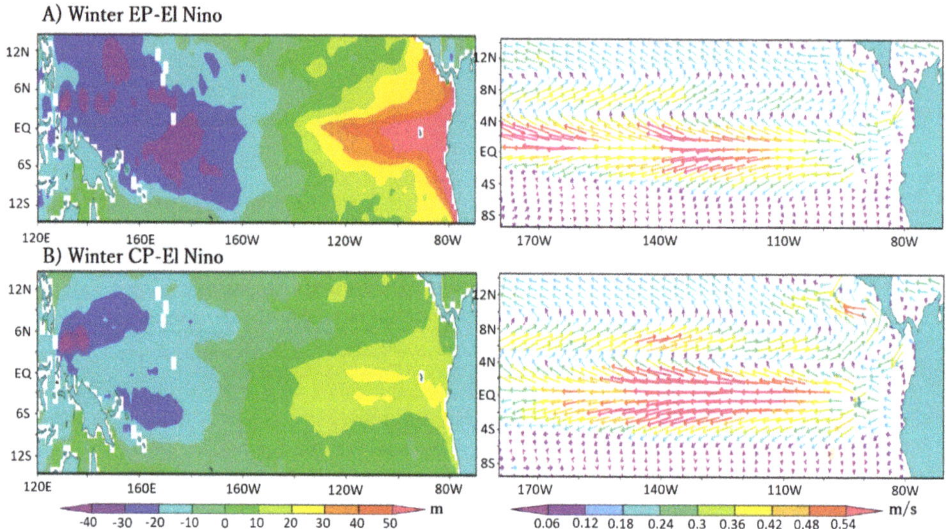

Figure 10. Same as Figure 9, but for winter from (**A**) the EP El Niño and (**B**) CP El Niño. Negative (Positive) D20 anomalies denote shoaling (deepening).

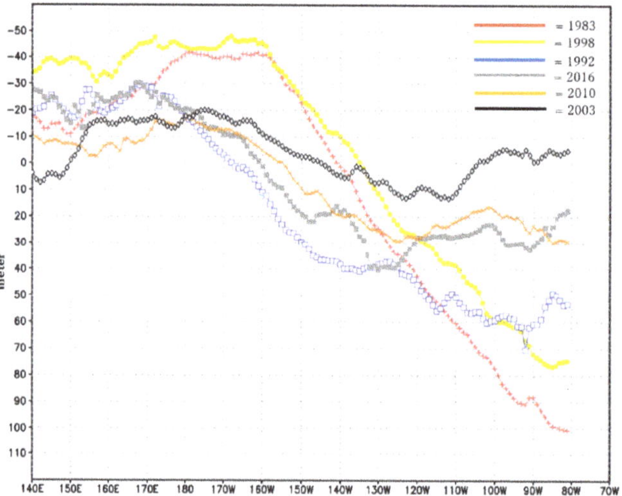

Figure 11. Equatorial Pacific D20 anomaly for January. Each type of El Niño was represented by two events; 1983 and 1998 representing EP type; 1992 and 2016 representing MIX type; 2003 and 2010 representing CP type.

With regard to the NECC's meridional shifting, the seasonal southward shift occurred between mid-winter to spring, then traveled to the north in the second half of the year. This result is consistent with Shin and Qiu [8]. We also found that the southward shift was missing in several El Niño years, specifically the CP and MIX El Niño. During these events, the absence of the southward shift in the CP and MIX El Niño agreed well with the SB pattern. From the SB analysis generated by local wind stress curl, the zero contour of the eastward transport near the equator was located between 0° and 1° S during the EP El Niño. Meanwhile, the zero contour migrated to the north of 3° N and 4° N for the

EP and MIX El Niño types, respectively. Prior studies, however, have shown the NECC's southward movement during the EP El Niño [8,12].

5. Conclusions

In the present paper, reanalysis data examined the NECC variations east of the dateline and how they responded to the three types of El Niño. Our findings led us to the following conclusions:

(1)   The NECC was considerably stronger during the EP and MIX El Niño than during the CP El Niño for the spring and summer, which are frequently linked with the development period. During the fall, the three types of El Niño did not differ significantly in terms of affecting the NECC's variations. These stronger and weaker NECCs east of the dateline are influenced by the thermocline depth oscillations across the Pacific Ocean.

(2)   Going through a mature phase, which is generally in the winter, the NECC during the EP El Niño was weaker than the NECC during the CP and MIX El Niño. This finding provides a new suggestion for future investigation because it has not been documented in previous studies.

(3)   The NECC shift to the south is absent during the CP and MIX El Niño events because of the local wind stress distribution. This southward shift, however, remained apparent during the EP El Niño.

Based on the results of this study, we give a fresh suggestion regarding the influence of the three types of El Niño on the NECC variations east of the dateline for each season. The NECC appeared to react differently in each season to the three types of El Nino. After thorough research, we discovered that the NECC has a propensity to steadily strengthen from the development to the mature phase of MIX El Niño events, which has not been accomplished before by previous research. Furthermore, the conclusions of this study enable further investigation into the behavior of other ocean circulations in the Pacific Ocean in response to the MIX El Niño event, for which little information is available.

Author Contributions: Conceptualization, Y.J.W. and U.J.W.; methodology, Y.J.W. and Y.H.; software, Y.J.W.; validation, Y.J.W. and Y.H.; formal analysis, Y.J.W. and Y.H.; investigation, Y.J.W. and Y.H.; resources, Y.J.W. and U.J.W.; data curation, Y.J.W. and U.J.W.; writing—original draft preparation, Y.J.W.; writing—review and editing, Y.J.W. and Y.H.; visualization, Y.J.W.; supervision, Y.H.; project administration, Y.H.; funding acquisition, Y.H. All authors have read and agreed to the published version of the manuscript.

Funding: This study was financially supported by a Grant-in-Aid for Scientific Research (C-2) from the Ministry of Education, Culture, Sports, Science, and Technology of Japan (20K04708).

Institutional Review Board Statement: Not applicable.

Informed Consent Statement: Not applicable.

Data Availability Statement: All of the study's data is available for free download at http://apdrc.soest.hawaii.edu/las/v6/dataset (accessed on 30 July 2021).

Acknowledgments: The first author at The University of the Ryukyus Department of Physics and Earth Sciences who is funded by the Japanese Ministry of Education, Culture, Sports, Science and Technology (MEXT).

Conflicts of Interest: The authors declare no conflict of interest.

References

1.   Zhao, J.; Li, Y.; Wang, F. Seasonal Variation of the Surface North Equatorial Countercurrent (NECC) in the Western Pacific Ocean. *Chin. J. Oceanol. Limnol.* **2016**, *34*, 1332–1346. [CrossRef]
2.   Wijaya, Y.J.; Hisaki, Y. Differences in the reaction of north equatorial countercurrent to the developing and mature phase of ENSO events in the western Pacific Ocean. *Climate* **2021**, *9*, 57. [CrossRef]
3.   Johnson, G.C.; Sloyan, B.M.; Kessler, W.S.; McTaggart, K.E. Direct measurements of upper ocean currents and water properties across the tropical Pacific during the 1990s. *Prog. Oceanogr.* **2002**, *52*, 31–61. [CrossRef]

4. Hsin, Y.-C. Trends of the pathways and intensities of surface equatorial current system in the north Pacific Ocean. *J. Clim.* **2016**, *29*, 6693–6710. [CrossRef]

5. Donguy, J.R.; Meyers, G. Mean annual variation of transport of major currents in the tropical Pacific Ocean. *Deep Sea Res. Part I Oceanogr. Res. Pap.* **1996**, *43*, 1105–1122. [CrossRef]

6. Webb, D.J. On the role of the North Equatorial Counter Current during a strong El Niño. *Ocean Sci.* **2017**, *14*, 633–660. [CrossRef]

7. Seo, I.; Lee, Y.I.; Kim, W.; Yoo, C.M.; Hyeong, K. Movement of the intertropical convergence zone during the mid-pleistocene transition and the response of atmospheric and surface ocean circulations in the central equatorial Pacific. *Geocherm. Geophys. Geosyst.* **2015**, *16*, 3973–3981. [CrossRef]

8. Hsin, Y.-C.; Qiu, B. The impact of Eastern-Pacific versus Central-Pacific El Niños on the North Equatorial Countercurrent in the Pacific Ocean. *J. Geophys. Res.* **2012**, *117*, C11017. [CrossRef]

9. Zhao, J.; Li, Y.; Wang, F. Dynamical responses of the west Pacific North Equatorial Countercurrent (NECC) system to El Niño events. *J. Geophys. Res. Ocean.* **2013**, *118*, 2828–2844. [CrossRef]

10. Wang, L.-C.; Wu, C.-R. Contrasting the Flow Patterns in the Equatorial Pacific between Two Types of El Niño. *Atmosphere-Ocean* **2013**, *51*, 60–74. [CrossRef]

11. Chen, X.; Qiu, B.; Du, Y.; Chen, S.; Qi, Y. Interannual and interdecadal variability of the North Equatorial Countercurrent in the Western Pacific. *J. Geophys. Res. Oceans.* **2016**, *121*, 7743–7758. [CrossRef]

12. Tan, S.; Zhou, H. The observed impacts of the two types of El Niño on the North Equatorial Countercurrent in the Pacific Ocean. *Geophys. Res. Lett.* **2018**, *45*, 10493–10500. [CrossRef]

13. Ashok, K.; Behera, S.K.; Rao, S.A.; Weng, H.; Yamagata, T. El Niño Modoki and its possible teleconnection. *J. Geophys. Res.* **2007**, *112*, C11007. [CrossRef]

14. Kug, J.-S.; Jin, F.-F.; An, S.-I. Two types of El Niño events: Cold tongue El Niño and warm pool El Niño. *J. Clim.* **2009**, *22*, 1499–1515. [CrossRef]

15. Yeh, S.-W.; Kug, J.-S.; Dewitte, B.; Kwon, M.-H.; Kirtman, B.-P.; Jin, F.-F. El Niño in a changing climate. *Nature* **2009**, *461*, 511–514. [CrossRef] [PubMed]

16. Kao, H.-Y.; Yu, J.-Y. Contrasting eastern-Pacific and central-Pacific types of ENSO. *J. Clim.* **2009**, *22*, 615–632. [CrossRef]

17. Hu, X.; Yang, S.; Cai, M. Contrasting the eastern Pacific El Niño and the central Pacific El Niño: Process-based feedback attribution. *Clim. Dyn.* **2016**, *47*, 2413–2424. [CrossRef]

18. Wen, N.; Liu, Z.; Li, L. Direct ENSO impact on East Asian summer precipitation in the developing summer. *Clim. Dyn.* **2018**, *52*, 6799–6815. [CrossRef]

19. Wu, B.; Zhou, T.; Li, T. Contrast of Rainfall-SST Relationship in the Western Pacific between the ENSO-Developing and ENSO Decaying Summers. *J. Clim.* **2009**, *22*, 4398–4405. [CrossRef]

20. Balmaseda, M.A.; Mogensen, K.; Weaver, A.T. Evaluation of the ECMWF ocean reanalysis system ORAS4. *Q. J. R. Meteorol. Soc.* **2013**, *139*, 1132–1161. [CrossRef]

21. Balmaseda, M.A.; Trenberth, K.E.; Källén, E. Distinctive climate signals in reanalysis of global ocean heat content. *Geophys. Res. Lett.* **2013**, *40*, 1754–1759. [CrossRef]

22. Zuo, H.; Balmaseda, M.A.; Mogensen, K.; Tietsche, S. *OCEAN5: The ECMWF Ocean Reanalysis System and Its Real-Time Analysis Component*; Technical Report 823; ECMWF: Reading, UK, 2018. [CrossRef]

23. Hersbach, H.; Bell, B.; Berrisford, P.; Hirahara, S.; Horányi, A.; Muñoz-Sabater, J.; Nicolas, J.; Peubey, C.; Radu, R.; Schepers, D.; et al. The ERA5 global reanalysis. *Q. J. R. Meteorol. Soc.* **2020**, *146*, 1999–2049. [CrossRef]

24. Molina, O.M.; Gutiérrez, C.; Sánchez, E. Comparison of ERA5 surface wind speed climatologies over Europe with observations from the HadISD dataset. *Int. J. Climatol.* **2021**, *41*, 4864–4878. [CrossRef]

25. Kok, P.H.; Mohd Akhir, M.F.; Tangang, F.; Husain, M.L. Spatiotemporal trends in the southwest monsoon wind-driven upwelling in the southwestern part of the South China Sea. *PLoS ONE* **2017**, *12*, e0171979. [CrossRef]

26. Kessler, W.S.; Johnson, G.C.; Moore, D.W. Sverdrup and Nonlinear Dynamics of the Pacific Equatorial Currents. *J. Phys. Oceanogr.* **2003**, *33*, 994–1008. [CrossRef]

27. Reynolds, R.W.; Rayner, N.A.; Smith, T.M.; Stokes, D.C.; Wang, W. An improved in situ and satellite SST analysis for Climate. *J. Clim.* **2002**, *15*, 1609–1625. [CrossRef]

28. Björnsson, H.; Venegas, S. *A Manual for EOF and SVD Analyses of Climate Data*; McGill University, Department of Atmospheric and Oceanic Sciences and Centre for Climate and Global Change Research Tech. Rep.: Montréal, QC, Canada, 1997; 52p.

29. Hannachi, A.; Jolliffe, I.T.; Stephenson, D.B. Empirical orthogonal functions and related techniques in atmospheric science: A review. *Int. J. Climatol.* **2007**, *27*, 1119–1152. [CrossRef]

30. Yu, J.-Y.; Zou, Y.; Kim, S.T.; Lee, T. The changing impact of El Niño on US winter temperatures. *Geophys. Res. Lett.* **2012**, *39*, L15702. [CrossRef]

31. Zhu, A.; Xu, H.; Deng, J.; Ma, J.; Li, S. El Niño-Southern Oscillation (ENSO) effect on interannual variability in spring aerosols over East Asia. *Atmos. Chem. Phys.* **2021**, *21*, 5919–5933. [CrossRef]

32. Paek, H.; Yu, J.-Y.; Qian, C. Why were the 2015/2016 and 1997/1998 extreme El Niños different? *Geophys. Res. Lett.* **2017**, *44*, 1848–1856. [CrossRef]

33. Zhang, Z.; Ren, B.; Zheng, J. A unified complex index to characterize two types of ENSO simultaneously. *Sci. Rep.* **2019**, *9*, 8373. [CrossRef] [PubMed]

34. Chen, M.; Li, T. Why 1986 El Niño and 2005 La Niña evolved different form a typical El Niño and La Niña. *Clim. Dyn.* **2018**, *51*, 4309–4327. [CrossRef]
35. Xu, K.; Huang, R.X.; Wang, W.; Zhu, C.; Lu, R. Thermocline fluctuations in the equatorial Pacific Related to the two types of El Niño event. *J. Clim.* **2017**, *30*, 6611–6627. [CrossRef]

Journal of
*Marine Science
and Engineering*

Article

General and Local Characteristics of Current Marine Heatwave in the Red Sea

Abdulhakim Bawadekji [1,*], Kareem Tonbol [2], Nejib Ghazouani [3], Nidhal Becheikh [3] and Mohamed Shaltout [4,*]

1   College of Science, Northern Border University, Arar 91431, Saudi Arabia
2   College of Maritime Transport and Technology, Arab Academy for Science, Technology and Maritime Transport, Alexandria 1209, Egypt; ktonbol@aast.edu
3   College of Engineering, Northern Border University, Arar 91431, Saudi Arabia; nejib.ghazouani@nbu.edu.sa (N.G.); Nidhal.Becheikh@nbu.edu.sa (N.B.)
4   Oceanography Department, Faculty of Science, Alexandria University, Alexandria 21526, Egypt
*   Correspondence: bawadekji@nbu.edu.sa (A.B.); mohamed.shaltot@alexu.edu.eg (M.S.); Tel.: +20-1-005-255-393 (M.S.)

Citation: Bawadekji, A.; Tonbol, K.; Ghazouani, N.; Becheikh, N.; Shaltout, M. General and Local Characteristics of Current Marine Heatwave in the Red Sea. *J. Mar. Sci. Eng.* **2021**, *9*, 1048. https://doi.org/10.3390/jmse9101048

Academic Editor: Francisco Pastor Guzman

Received: 19 August 2021
Accepted: 16 September 2021
Published: 23 September 2021

Publisher's Note: MDPI stays neutral with regard to jurisdictional claims in published maps and institutional affiliations.

Abstract: In the ocean, heat waves are vital climatic extremes that can destroy the ecosystem together with ensuing socioeconomic consequences. Marine heat waves (MHW) recently attracted public interest, as well as scientific researchers, which motivates us to analyze the current heat wave events over the Red Sea and its surrounding sea region (Gulf of Aden). First, a comprehensive evaluation of how the extreme Red Sea surface temperature has been changing is presented using 0.25° daily gridded optimum interpolation sea surface temperature (OISST, V2.1) data from 1982 to 2020. Second, an analysis of the MHW's general behavior using four different metrics over the study area, together with a study of the role of climate variability in MHW characteristics, is presented. Finally, the main spatiotemporal characteristics of MHWs were analyzed based on three different metrics to describe MHW's local features. Over the studied 39 years, the current results showed that the threshold of warm extreme sea surface temperature events (90th percentile) is 30.03 °C, providing an additional average thermal restriction to MHW threshold values (this value is changed from one grid to another). The current analysis discovered 28 separate MHW events over the Red+, extending from 1988 to 2020, with the four longest events being chosen as a study case for future investigation. For the effect of climate variability, our results during the chosen study cases prove that ENSO and ISMI do not play a significant role in controlling MHW characteristics (except the MHW intensity, which has a clear relation with ENSO/ISMI) on Red+. Moreover, the chlorophyll concentration decreases more significantly than its climatic values during MHW events, showing the importance of the MHW effect on biological Red Sea features. In general, the MHW intensity and duration exhibit a meridional gradient, which increases from north to south over the Red Sea, unlike the MHW frequency, which decreases meridionally.

Keywords: marine heat wave; climate variability; sea surface temperature; extreme events; global climate models; ecosystems

1. Introduction

Several notable marine heat waves (MHWs) events occurred globally—long periods of abnormal sea surface temperature extremes have had severe impacts on marine ecosystems, as stated by [1]. According to Sparnocchia et al. [2], Oliver et al. [3] and Holbrook et al. [4], MHWs are driven by a range of physical mechanisms, such as air–sea heat fluxes that coincide with atmospheric heat waves and/or horizontal temperature advection. Regardless of the mechanisms that drive individual MHWs, there is a growing acceptance that anthropogenic climate change has raised the likelihood of recent MHWs dramatically, including the following occasions. Prominent occasions occurred over the Red Sea, especially: in its northern basin, as stated by Chaidez et al. [5]; in the Gulf of Aqaba, as stated by Shaltout [6];

99

along the Mediterranean Sea including the central Ligurian Sea (Sparnocchia et al. [2]), the central basin (Olita et al. [7]) and the eastern basin (Ibrahim et al. [8]); over the eastern Indian ocean, especially along the Western Australian coast (Pearce and Feng [9]) and across northern Australia (Benthuysen et al. [10]); over the northeastern Pacific ocean (Bond et al. [11]); over the northwestern Atlantic ocean (Chen et al. [12]). These occasions resulted in significant environmental and financial impacts, including a reduced chlorophyll-a concentration (Bond et al. [11]), continuous coral bleaching (Hughes et al. [13]), the death of fish (Caputi et al. [14]), mass mortality (Garrabou et al. [15]), geographical and seasonal shifts of marine species (Mills et al. [16]; Cavole et al. [17]) and economic problems (Mills et al. [16]). Considering the current/projected warming trends over the Red Sea (Shaltout [6]), as well as the potential for deep ecological and social consequences, assessing MHWs patterns and trends is currently a critical topic concerning the Red Sea.

The Red Sea surface temperature (SST) experiences a current warming (1982–2017) trend of 0.29 °C/decade, while the annual mean SST over the Red Sea during the current century is expected to increase by 0.6–3.2 °C relative to the 2006–2035 period, as stated by Shaltout [6]. Based on the SST analyses over the Red Sea, Shaltout [6] indicated that the Red Sea suffers heat wave events that currently occur during approximately 3% of each year ($\cong$10 days annually), which is expected to be more frequent by the end of the 21st century. On the other hand, Bindoff et al. [18] showed that changes in extreme weather events affected largely different species in comparison to the effect of changes in mean conditions, indicating the importance of the evaluation of MHWs events rather than SST trends, especially over the warming climate regions (e.g., the Red Sea climate). Moreover, marine organisms in the Red Sea, which is a semi-enclosed basin, are unable to migrate north. Thus, heat waves have been linked to some of the most catastrophic environmental changes over the Red Sea (Hodgkinson et al. [19]; Chaidez et al. [5]).

Some earlier relevant studies concerning MHWs' duration, frequency and intensity are available. According to Oliver et al. [3], there has been a significant global increase in MHWs' frequency (duration), which has increased by 34% (17%) from 1925 to 2016. In fact, over 18 days, from 29 July 2012 to 15 August 2012, the longest-detected MHWs occurred in the Gulf of Aqaba, which is north of the Red Sea (Shaltout [6]). In terms of the overall number of events, as well as the intensity and duration, Genevier et al. [20] confirmed that MHWs had a distinct spatial pattern in the Red Sea. The southernmost tip of the Red Sea and the eastern coast of the northern region had the greatest number of days of MHWs, whereas the western coast of the southern region had the most intense events, and the most persistent events occurred over the eastern coast of the southern Red Sea.

In general, this increase in the MHWs events over the Red Sea emphasizes the urgent necessity to describe the MHWs' main characteristics and their link to climatic variability, together with identifying the regions in the Red Sea that are vulnerable to MHWs.

In the current study, MHWs were carefully analyzed to find out their characteristics over the Red Sea, together with the Gulf of Aden (hereafter, Red+; Figure 1). As such, understanding the MHWs' variabilities in the study region is the main aim of this study, in order to be able to implement appropriate early awareness procedures related to the thermal stress on various marine sectors (e.g., coral reef bleaching), together with finding suitable regional climate policies to cope with climatic change issues. The data used and methods are presented in Section 2. In Section 3, the results are included, while the summary and conclusion are covered in Section 4.

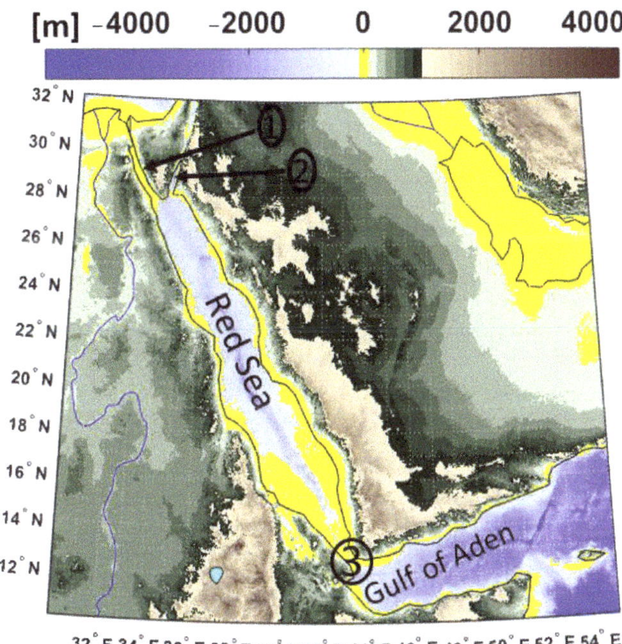

Figure 1. Digital elevation data of the Red Sea (data acquired from a global 30 arc-second interval grid (GEBCO: https://www.gebco.net/data_and_products/gridded_bathymetry_data/ [accessed on 10 May 2021]). The Gulfs of Suez (1), and Aqaba (2) together with the strait of Bab al Mandab were shown in the figure.

2. Data and Methods

In general, the MHWs are described by their frequency, duration and intensity. Their definition will have an impact on the analyses of the magnitude and duration of such events. According to Hobday et al. [1], MHWs originate when SSTs exceed a seasonally changeable threshold, defined as the 90th percentile of climatic SST mean for at least 5 consecutive days. In addition, Chaidez et al. [5] used a new definition based on considering yearly maximum SST above the climatic maximum SST by a given threshold chosen at 0.25 °C intervals between 0.5 and 1.5 °C as a base to define MHWs events. Finally, Darmaraki et al. [21] used the climatological 99th percentile threshold, based on daily SST over the period 1976 to 2005.

2.1. Data Used

Gridded daily averaged SST fields on a 0.25° horizontal resolution were obtained from NOAA optimum interpolation sea surface temperature data (OISST; version 2) over a 39-year period from January 1982 to December 2020. On a regular global grid, the OISST combines satellite ocean skin temperatures with data from in situ platforms (ships and buoys), as stated by Reynolds et al. [22] and Reynolds [23]. The OISST products, according to Banzon et al. [24], do not capture diurnal variations and do not represent a specific time of day because they are made up of data collected throughout the day. The International Comprehensive Ocean-Atmosphere Data Set (ICOADS) provided the in situ platform measurements for the OISST products (Worley et al. [25]. Karnauskas and Jones [26] highlighted the higher density of the in situ measurements in the ICOADS data bank, especially in the Red Sea. Thus, the OISST products are a relevant tool to study local features of the Red+. Moreover, the OISST data feasibility in describing SST over the Red+ has been carried out by inter-comparisons with independent in situ measurements (Shaltout [6]), confirming the excellent agreement between OISST and in situ measurements. The OISST products are freely available as gridded NetCDF (network Common Data Form) via HTTP link (https://www.ncei.noaa.gov/data/sea-surface-temperature-optimum-interpolation/v2.1/access/avhrr/ [accessed on 1 February 2021]). These data will be used for determining the climatological SST mean and 90th percentile of the historical SST distribution for each day of the year.

Chlorophyll concentration in sea water (chlor_a). Gridded daily data on chlor_a concentration was obtained from the MODIS (Moderate Resolution Imaging Spectroradiometer) sensor database on the NASA Aqua satellite. Currently, the Level 3 standard mapped image with 0.04° (~4-km) resolution was used to study the variability of chlor_a concentration over Red+. SeaWiFS Mission webpage [27] provides detailed description about chlor_a (validations and documentation). These data have been used largely over the Red Sea (Brewin et al. [28]; Eladawy et al. [29]; Shaltout [6]). These data will be used to understand the effect of MHW on the chlor_a concentration, especially during the selected study cases.

El Niño/Southern Oscillation (ENSO) index. The time series of the leading combined empirical orthogonal function (EOF) of five different variables (SST, sea level pressure, outgoing longwave radiation and zonal and meridional components of the surface wind) over the tropical Pacific basin (30° S–30° N and 100° E–70° W) is the bi-monthly multivariate ENSO index (MEI.v2). More detailed information about MEI.v2 is available at Zhang et al. [30]. MEI.v2 data were extracted from the physical science laboratory (https://psl.noaa.gov/enso/mei/ [accessed on 23 May 2021]) for 12 overlapping bi-monthly "seasons" (December–January, January–February,..., November–December) in order to both decrease the impact of higher frequency intra-seasonal variability and take into account ENSO's seasonality. El Niño periods and La Niña periods were identified based on a threshold of ±0.5. These data are used to study the climate variability role in the characteristics of MHWs.

Indian Summer Monsoon Index (ISMI), which contributes to wind circulation and temperature distribution, lasts only during the summer season (June to September). Monthly ISMI data were obtained from the University of Hawaii Data Center (http://apdrc.soest.hawaii.edu/projects/monsoon/seasonal-monidx.html [accessed on 10 June 2021]) over 1982–2019. According to Wang et al. [31], the ISMI is an 850-hPa zonal wind difference between a southern zone (40–80° E, 5–15° N) and a northern region (70–90° E, 20–30° N). ISMI is used to study how the MHWs' characteristics are related to ISMI.

Sea surface radiation budget components, including surface latent heat flux (SLHF), surface sensible heat flux (SSHF), surface net thermal radiation (SNTR) and surface net solar radiation (SNSR), were obtained from ERA5 reanalysis database during 1982–2020 (https://cds.climate.copernicus.eu/cdsapp#!/dataset/reanalysis-era5-single-levels?tab=form [accessed on 10 June 2021]). ERA5, which replaced the successful previous version of ERA-Interim, featured a highly spatial/temporal grid point, as well as major improvements in core dynamics and model physics [32]. By merging weather model data with observational data from ground sensors and satellites, ERA5 provides an accurate long-term record of global climate and weather [33]. Furthermore, the total heat loss to the atmosphere (Floss) equals SLHF + SLHF + SNTR +SNSR (fluxes are positive when directed away from surface to the atmosphere).

2.2. Extreme Red+ Sea Surface Temperature

Daily means in a 39-year period SST data were used to determine the thresholds of extreme sea surface temperature events (hereafter, ETEs) over the Red+; thus, SST values above the 90th percentile were considered as warm ETEs and SST values below 10th percentile were considered as cold ETEs. The frequency of warm ETE ($= \frac{number\ of\ extreme\ warm\ seasons}{total\ number\ of\ seasons} \times 100$) and the magnitude of warm ETE (=The maximum values of seasonally averaged SST) were used to identify the warm extreme Red+ sea surface temperature.

2.3. Analyses of the Marine Heat Waves over the Red+

Daily OISST SST data were used to identify MHWs, and a set of four metrics were developed to quantify MHWs: (1) MHWs categories, (2) frequency (the number of individual MHWs events that occur annually), (3) duration (the number of MHWs days over the entire 39-year study period) and (4) intensity of an event (mean, maximum and accumulative). Mean (maximum) intensity is described as the average (maximum) SST ($^{\circ}$C) above the climatological mean during an event, whereas the cumulative intensity ($^{\circ}$C-day) is described as the mean intensity multiplied by the event's duration.

The 90th percentile of the historical SST distribution through a baseline period over the Red+ is used to identify MHWs categories (Hobday et al. [1]; Hobday et al. [34]). Thus, MHWs categories were defined as the local difference between climatological 90th percentile and climatological mean. According to Hobday et al. [34], different MHWs categories are defined by a magnitude scale calculated based on the multiples of this local difference. Moderate MHWs category falls in the range (1–2 * local difference), strong MHWs category falls in the range (2–3 * local difference), severe MHWs category falls in the range (3–4 * local difference), extreme MHWs category falls in the range (4–5 * local difference).

A baseline period of 39 years (1982–2020) was used to identify the changeable climatological SST mean and 90th percentile threshold based on each day of the year. The climatological SST mean for a specific day was calculated by averaging the daily SST values within an 11-day window ranging from 5 days prior to the specific day to 5 days after the specific day over the entire 39-year baseline period. The minimum duration of an MHW event was set to be five days (Hobday et al. [1]), and an intermittent period of up to 2 consecutive days or fewer (if the MHWb duration>10 days) was considered to be a single MHW event. Thresholds of warm ETEs ($\approx$30 $^{\circ}$C over the Red+ and changeable from grid to grid) were used as an additional thermal restriction to MHWs threshold values.

General characteristics of MHWs categorization over the Red+ were analyzed every year to report the most important MHWs study cases. Local features of MHWs were furtherly analyzed over the selected study cases that have the longest duration.

2.3.1. The Role of Climate Variability

The annual effects of the ENSO—the most reliable indicator of global climate change—on the MHWs characteristics were studied by comparing their annual patterns. Moreover, the annual effect of ISMI on the MHWs variabilities was also studied.

2.3.2. The Role of the Sea Surface Radiation Budget Components

A direct comparison between MHW intensity and different sea surface radiation budget components (SLHF, SSHF, SNTR, SNSR and F_{loss}) was carried out using the correlation coefficient (R) and a number of observations (n) during the selected study cases. Furthermore, a direct comparison between MHWs' intensity and sea surface radiation budget for different components over the study period was carried out to assess the role of the sea surface radiation budget on MHW.

2.4. Main Spatiotemporal Characteristics of MHWs over the Red+

The spatial and temporal variability of MHWs characteristics were analyzed in Red+ over a studied 39-year period, focusing on annual variability. Annual averaged MHWs' intensity ($^{\circ}$C), marine heat wave duration over the 39-year study period (day) and marine heat wave frequency (event per year) were selected to characterize the main spatiotemporal features of MHWs.

3. Results

3.1. Extreme Red+ Sea Surface Temperature

Based on a 39-year period, the time series of the seasonal mean SST (Figure 2) over the Red+ showed that the threshold of warm ETEs (90th percentile) is 30.03 $^{\circ}$C and cold ETEs

(10th percentile) is 25.13 °C. The frequency of warm ETEs is 4.5% and is most pronounced after 1998, particularly in the summer of 1998, 2017, 2015, 2002, 2001, 2019 and 2020 (years arranged in ascending order). However, the magnitude of warm ETEs is 30.29 °C (occurring during the summer of 2020), partly due to the effects of climate change.

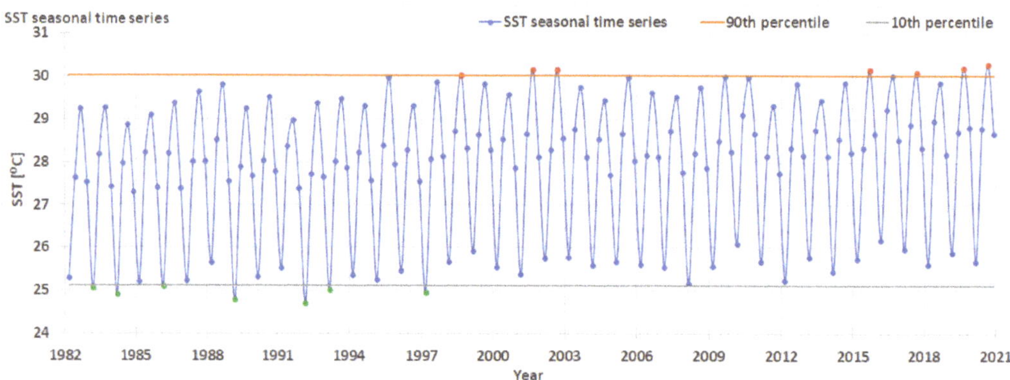

Figure 2. Time series of seasonally mean sea surface temperature over Red+. Warm ETEs (90th percentile) and cold ETEs (10th percentile) are presented in red and green, respectively.

The spatial variation in the extreme Red+ sea surface temperature over the years 1982 to 2020 is described in terms of the threshold and magnitude. The spatial distribution of thresholds of warm ETEs (Figure 3a) increased meridionally over the Red Sea, from the north (≈27 °C; the Gulfs of Suez and Aqaba) to the south (≈32 °C; western–northern part of the Red Sea), partly due to the amount of absorbed solar energy, together with the surface sea water circulation (Shaltout [6]). Moreover, thresholds of warm ETE are much higher in the northern part of the strait of Bab al Mandab in comparison to its southern part, partly due to the moderate SST effect in the Gulf of Aden by water exchange with the Indian Ocean. In the other direction, spatial distribution in the magnitude of warm ETEs (Figure 3b) values increased meridionally over the Red Sea, from the north (≈29 °C; the Gulfs of Suez and Aqaba) to the south (≈35 °C; southwestern coast of the Red Sea from 15–17° N), partly due to the surface sea water circulation (Shaltout [6]). Over the Gulf of Aden, the magnitude of warm ETEs showed a non-significant spatial range around 33 °C.

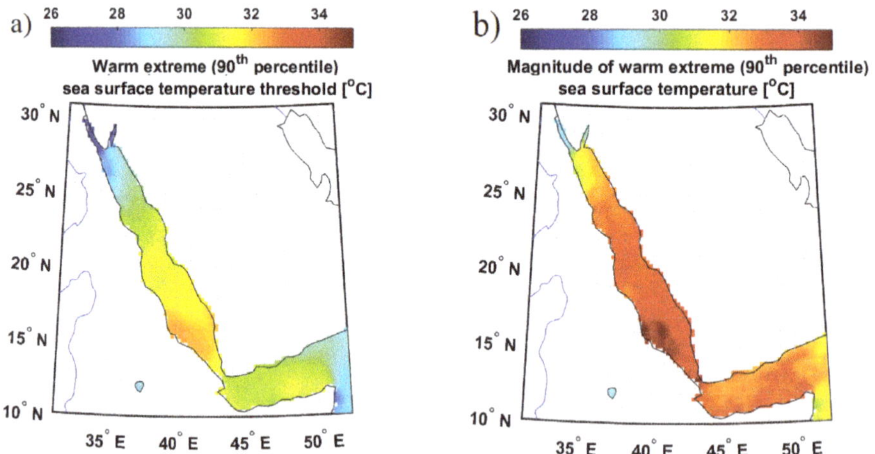

Figure 3. Spatial distribution of the threshold (**a**) and magnitude (**b**) of warm SST ETEs (90th percentile) over Red+.

3.2. Marine Heat Waves over the Red+

A categorization diagram of MHWs over the Red+ was drawn for every year to monitor the progress of MHWs, with the time as a general feature (Figure 4 and Table 1). For the years 1982–1986, there were no observed MHW events over the Red+. Over the years 1987–1991, there is only one MHW event, which extends 8 days, from 6–13 July/1988 (moderate MHWs category), showing the 1st MHW event for the Red+. Moreover, there is no observed MHW event over 1992–1996. From 1997 to 2001, three moderate MHW events (total number of MHWs = 60 days) were identified; the first event extends 9 days from 18 to 26 June 1997, the second event extends 12 days from 9 to 20 September 1998 and the third event extends 44 days (39 days without gaps) from 10 July to 22 August 2001. MHW events increased to seven moderate MHW events over the period 2002 to 2006 (total number of MHWs = 63 days). MHW events decreased to three moderate MHW events over the period 2007 to 2011 (total number of MHWs = 53 days). There are a total number of 66 days of MHWs in general over the Red+ during the 2012–2016 years, which are divided into six MHW moderate events. The total number of general Red+ MHWs is nearly double the previous periods (=128 days) over 2017–2020, and is divided into eight MH events (seven moderate MHW events and one strong MHW event). Generally, over the Red+ (1992–2020), the general MHW frequency is 0.72 event, and the MHW duration is 378 days. From the reported study cases (Table 1), the average duration of an MHW event is 13.5 ± 11.6 days; however, the average values of the maximum (mean) intensity of an event are 0.97 ± 0.23 (0.8 ± 0.1) °C. The average value of the accumulative intensity of an event is 11.52 ± 12.56 °C-day.

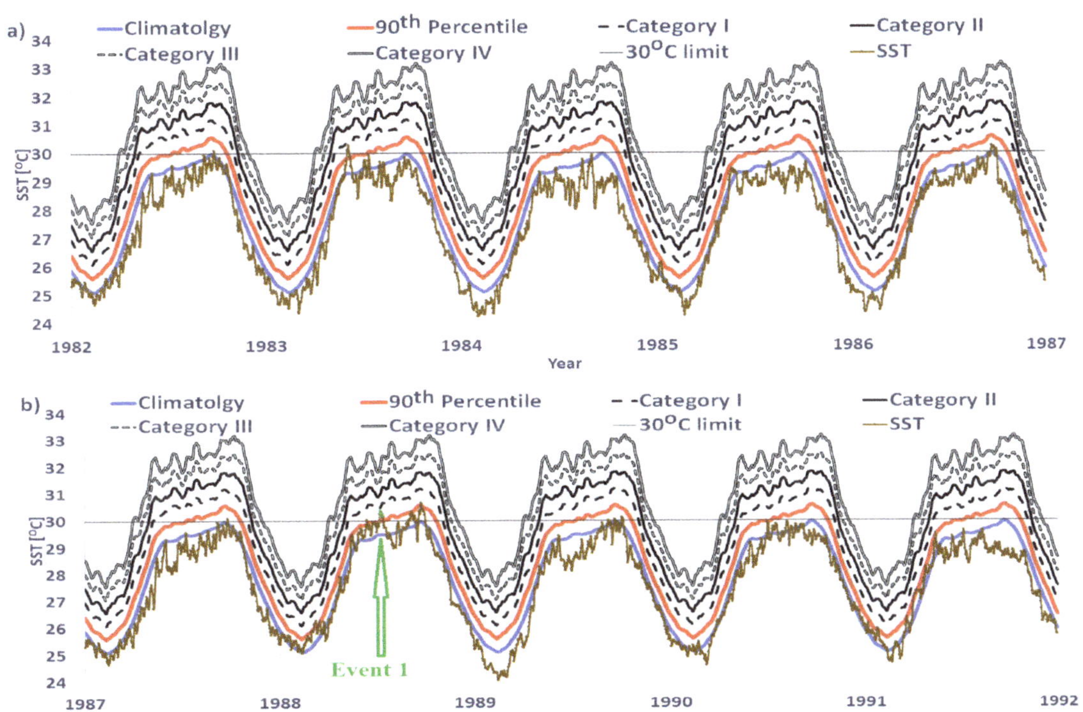

Figure 4. *Cont.*

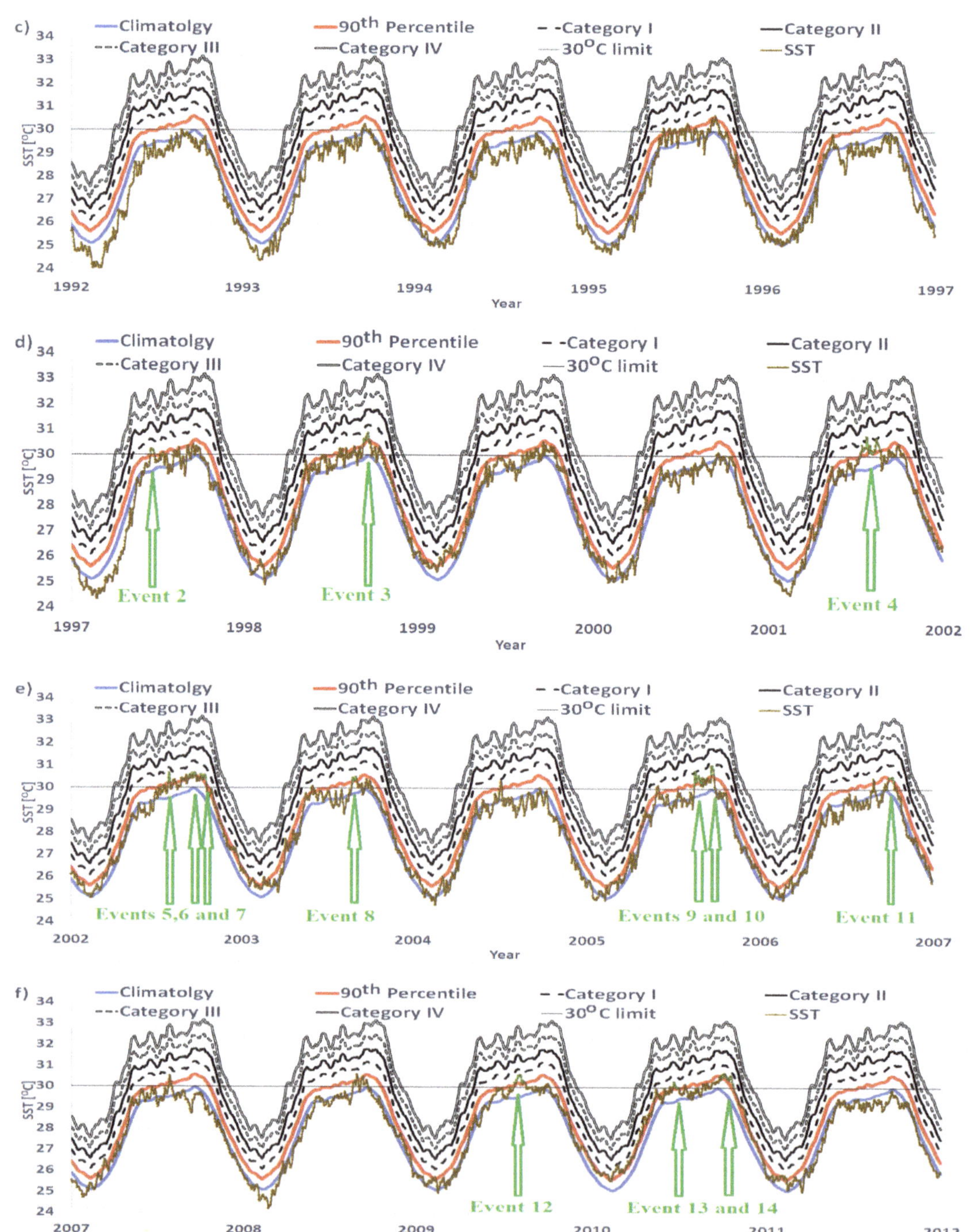

Figure 4. *Cont.*

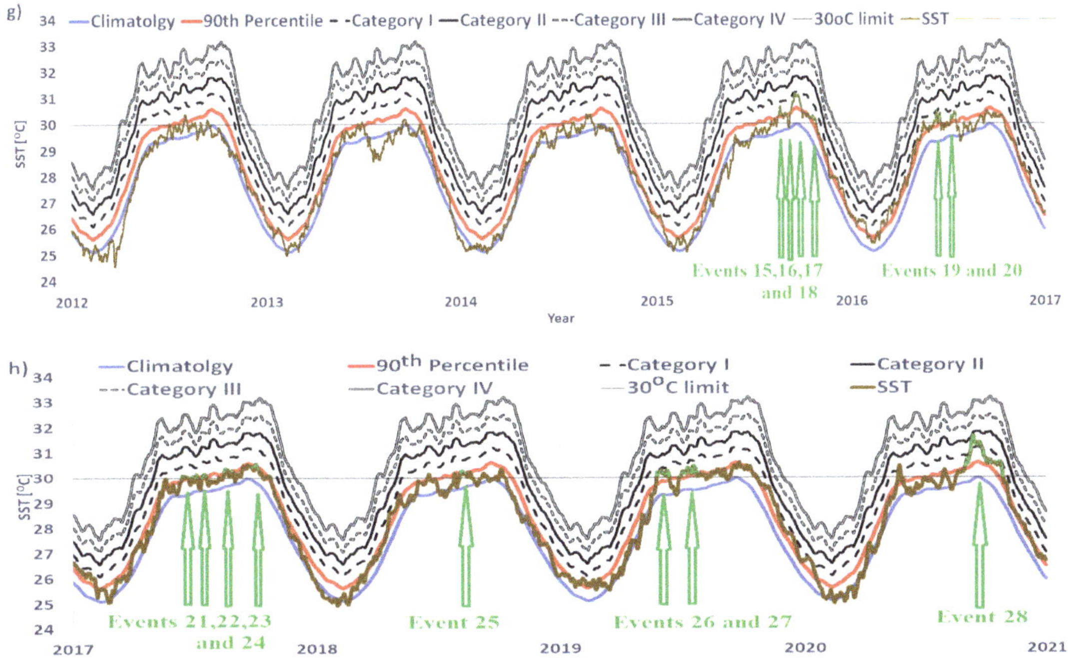

Figure 4. Marine heat waves over the Red+: a categorization diagram showing the observed temperature time series (dashed line) for each year from 1982 to 2020 (1982–1986 (**a**), 1987–1991 (**b**), 1992–1996 (**c**), 1997–2001 (**d**), 2002–2006 (**e**), 2007–2011 (**f**), 2012–2016 (**g**), and 2017–2020 (**h**)), the long-term regional climatology and the 90th percentile climatology together with the four categories.

Table 1. The characteristics of the recent MHWs over the Red+. The entry for each event lists date of peak intensity, the total duration of the event (first day, last day and total days), the category, maximum intensity (I_{max}, °C above the climatological mean) on that date, mean intensity (I_{mean}) along event's duration, the cumulative intensity ($I_{cum} = I_{mean} *$ event duration) and the proportion (p) of time spent in each of the four MHW categories along event's duration. As the presence of "gap days" connecting successive events, the proportions do not always sum up to 100%.

Event Number.	Date of Peak Intensity	First Day	Last Day	without Gaps Total Days	Category	I_{max} (°C)	I_{mean} (°C)	I_{cum} (°C-day)	Moderate (M)	Strong (Sg)	Severe (Sv)	Extreme (E)
1	8 July 1988	6 July 1988	13 July 1988	8	M	0.88	0.72	5.76	100	-	-	-
2	20 June 1997	18 June 1997	26 June 1997	9	M	0.90	0.82	7.45	100	-	-	-
3	19 September 1998	9 September 1998	20 September 1998	12	M	0.91	0.77	9.21	100	-	-	-
4	22 July 2001	10 July 2001	22 August 2001	39	M	1.30	0.82	36.03	89	-	-	-
5	31 July 2002	29 July 2002	2 August 2002	5	M	1.13	0.91	4.58	100	-	-	-
6	11 August 2002	9 August 2002	14 August 2002	6	M	0.74	0.68	4.1	100	-	-	-
7	15 October 2002	5 October 2002	16 October 2002	10	M	1.04	0.85	4.26	85	-	-	-
8	27 August 2003	26 August 2003	4 September 2003	10	M	0.73	0.60	5.14	100	-	-	-
9	16 August 2005	15 August 2005	31 September 2005	15	M	1.11	0.79	11.93	89	-	-	-

Table 1. *Cont.*

| Event Number. | Date of Peak Intensity | The Total Duration of the Event | | without Gaps Total Days | Category | Intensity (Calculated above the Climatological Mean) | | | P (%) | | | |
		First Day	Last Day			I_{max} (°C)	I_{mean} (°C)	I_{cun} (°C-day)	Moderate (M)	Strong (Sg)	Severe (Sv)	Extreme (E)
10	22 September 2005	19 September 2005	25 September 2005	7	M	1.04	0.91	6.37	100	-	-	-
11	7 October 2006	29 September 2006	8 October 2006	10	M	0.81	0.70	7.08	100	-	-	-
12	5 August 2009	23 July 2009	13 Aug2009	21	M	1.05	0.81	17.7	96	-	-	-
13	21 June 2010	18 June 2010	24 June 2010	7	M	0.93	0.82	5.74	100	-	-	-
14	18 October 2010	29 September 2010	23 October 2010	25	M	1.15	0.87	21.6	100	-	-	-
15	21 August 2015	18 August 2015	23 August 2015	6	M	0.96	0.78	4.66	100	-	-	-
16	31 August 2015	29 August 2015	2 September 2015	5	M	0.61	0.54	2.68	100	-	-	-
17	18 September 2015	9 September 2015	29 September 2015	21	M	1.18	0.94	19.77	100	-	-	-
18	24 October 2015	17 October 2015	26 October 2015	10	M	1.04	0.94	9.38	100	-	-	-
19	10 June 2016	6 June 2016	18 June 2016	12	M	1.19	0.90	10.97	92	-	-	-
20	14 July 2016	7 July 2016	18 July 2016	12	M	0.92	0.78	9.31	100	-	-	-
21	18 June 2017	17 June 2017	21 June 2017	5	M	0.75	0.72	3.60	100	-	-	-
22	16 July 2017	14 July 2017	18 July 2017	5	M	0.66	0.63	3.11	100	-	-	-
23	17 August 2017	15 August 2017	20 August 2017	6	M	0.71	0.66	3.97	100	-	-	-
24	22 October 2017	19 October 2017	23 October 2017	5	M	0.86	0.80	4.01	100	-	-	-
25	6 August 2018	5 August 2018	9 August 2018	5	M	0.72	0.66	3.30	100	-	-	-
26	25 May 2019	24 May 2019	11 June 2019	19	M	1.05	0.91	17.28	100	-	-	-
27	18 July 2019	27 June 2019	22 July 2019	26	M	0.97	0.77	19.90	100	-	-	-
28	13 September 2020	31 August 2020	26 October 2020	57	Sg	1.76	1.11	63.78	69	31	-	-

Generally, there are 28 MHW events over the Red+, and the longest four events were select for further investigations as a study case (Table 1; shaded events). As a regional average over Red+, MHW described a significant increasing trend of 0.49 annual events per decade (Figure 5a). The linear trend of the MHW annual duration in a regional average showed a significant increase of 7.5 days per decade, as seen in Figure 5b. Similarly, the averaged MHW intensity increased significantly (linear trend = 0.18 °C per decade), as seen in Figure 5c. On the other hand, the Red+ averaged SST showed a significant trend of 0.22 °C per decade (Figure 5d). The Red+ averaged MHW property time series has a clear interannual variability, with an average value of 1.38 events annually and an average intensity of 0.8 °C, together with an average duration of 22.5 days for an event. Moreover, the Red+ annual average SST is in a positive significant correlation with the annual averaged MHWs intensity (R = 0.75, *n* = 39), annual average MHWs duration (R = 0.72, *n* = 39) and annual MHWs frequency (R = 0.69, *n* = 39).

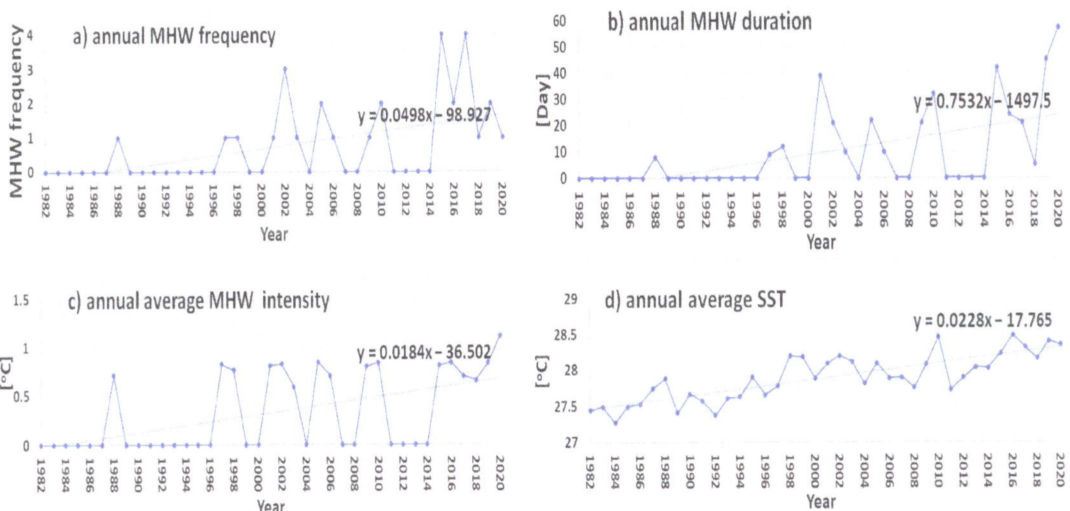

Figure 5. Sea surface temperature and marine heat wave properties over Red+.

The role of climate variability. The studied MHWs property time series (Figure 5) depict a significant interannual fluctuation over Red+. The scientific method of quantifying such a fluctuation is to examine the ENSO climatic states with the annual averaged MHWs intensity, annual averaged MHWs duration and MHWs frequency. The higher annual averaged MHWs intensity values are related to El Niño periods over most of the events (Figure 6a). The only event of El Niño that is not related to a higher intensity is 1983. The relationship between the frequency/duration and ENSO events was less clear than the intensity was (Figure 6c). The MHWs duration tends to increase (decrease) during El Niño (La Niña periods) periods, except during 1983 (years of the El Niño event), which correlated with no MHWs events, and during 2001 (year of La Niña periods), which correlated with a higher MHWs duration value. Similarly, the MHWs frequency tends to increase (decrease) during El Niño (La Niña periods) periods, except during 1983 and 2015.

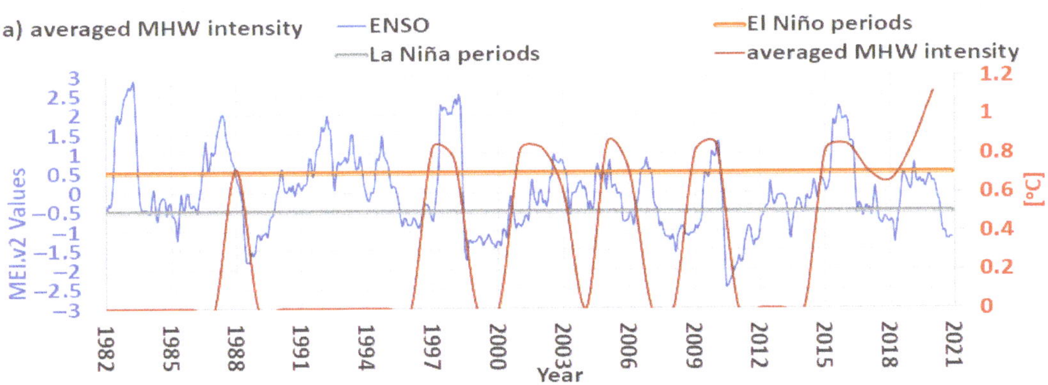

Figure 6. *Cont.*

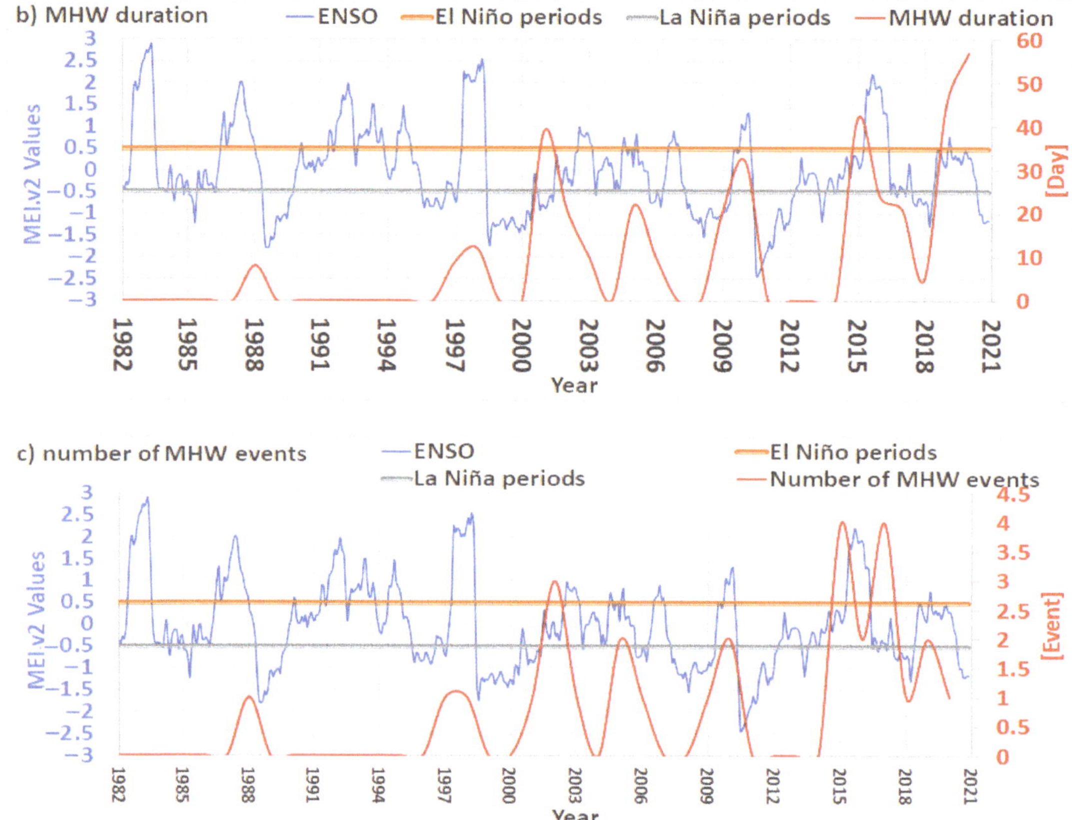

Figure 6. Annual time series of the bi-monthly multivariate El Niño/Southern Oscillation (ENSO) index (MEI.v2) over 1982–2020 are shown in primary axis. Moreover, the lower (higher) limit of El Niño (La Niña) periods is shown in orange (gray) color. In the secondary axis and in red color, intensity, duration and frequency were shown in (**a–c**), respectively.

On the other hand, the ISMI's relationship with the MHWs intensity is varied from one year to another over Red+. The higher (lower) values of the MHWs intensity during 1988, 2005, 2010 and 2016 (2018) are related to the higher (lower) values of the ISMI. At the same time, the higher values of the MHWs intensity during 1997, 2002 and 2015 occurred simultaneously with the lower ISMI values (Figure 7a). The ISMI rarely explains the higher values of the MHWS duration and frequency, such as during 2010 (Figure 7b,c).

The relationship between the intensity and ENSO/ISMI is clearer than the relationship between the frequency/duration wand ENSO/ISMI. Thus, there is another climatic mechanism that, if coupled with the ISMI and ENSO, will describe the MHWs' characteristics more clearly over Red+.

In statistical details, MEI.v2 and different MHW characteristics during the MHW events had an insignificant correlation at the 95% significance level (Figure 8). In the same, context, the ISMI described an insignificant correlation with the MHW annual average intensity, annual number of MHW events and annual MHW duration at the 95% significance level. At the 90% significance level, the ISMI showed a significant correlation only with the annual number of MHW events and annual MHW duration (Figure 9).

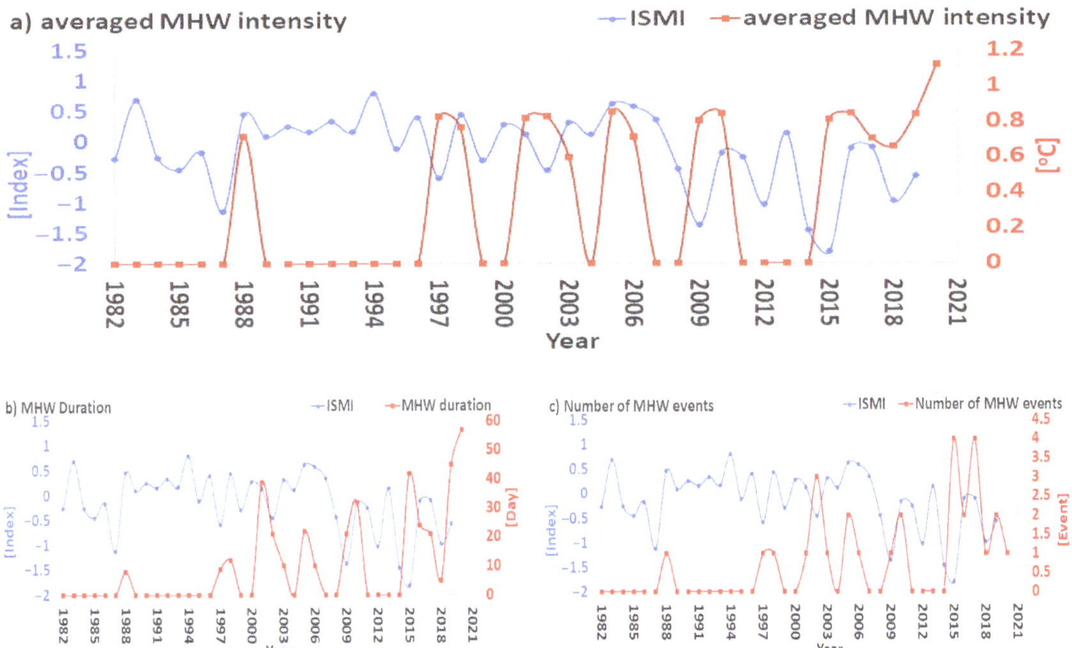

Figure 7. Annual time series of the Indian Summer Monsoon Index (ISMI) over 1982–2019 are shown in primary axis. In the secondary axis and in red color, intensity, duration and frequency are shown in (**a**–**c**), respectively.

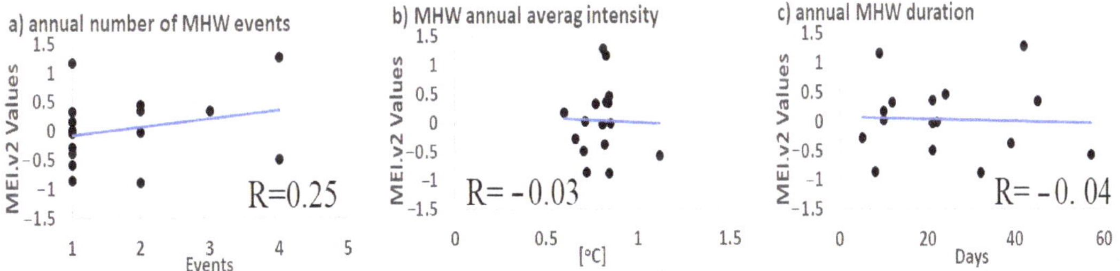

Figure 8. Scatter plots of bi-monthly multivariate ENSO index (MEI.v2) and annual number of MHW events (**a**), MHW annual average intensity (**b**) and annual MHW duration (**c**) during MHW events.

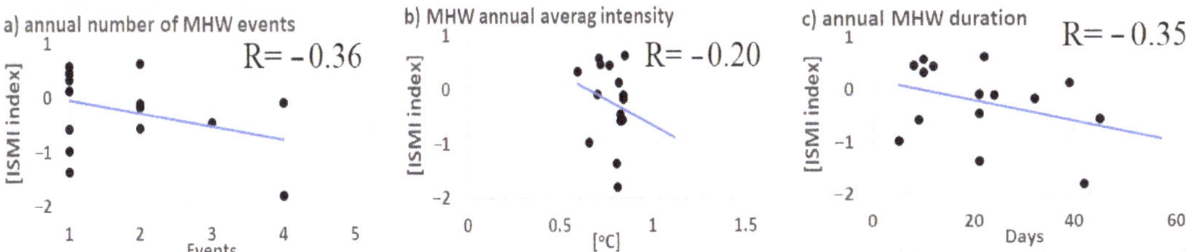

Figure 9. Scatter plots of Indian Summer Monsoon Index (ISMI) and annual number of MHW events (**a**), MHW annual average intensity (**b**) and annual MHW duration (**c**) during MHW events.

The role of the sea surface radiation budget components. The cross-correlation between different MHW characteristics and different components of the sea surface radiation budget (data not shown) showed insignificant values among each other on an annual level. This may indicate the need for further investigations based on the comparison between the range of the sea surface radiation budget components over the study period and during marine heat wave events (Table 2). Critical conditions of the occurrence of MHW events are less than 206 Wm^{-2} of SLHF going to the atmosphere, less than 13 Wm^{-2} of SSHF going to the atmosphere, less than 101 Wm^{-2} of SNTR going to the atmosphere and more than 210 Wm^{-2} of SNSR going to the ocean, as seen in Table 2.

Table 2. Range of the sea surface radiation budget components over the study period and during marine heat wave events (all values are in W m^{-2}).

	SLHF	SSHF	SNTR	SNSR	F_{loss}
Range over the study period	41 to 287	−14 to 90	50 to 163	−294 to −99	−152 to 287
Range during the marine heat wave events	57 to 206	−10 to 13	56 to 101	−279 to −210	−133 to −7

3.3. Selected Study Cases of Marine Heat Waves over the Red+

The first study case (peaked on 22 July 2001) centered around the center of the Red Sea (22.125° N and 38.375° E) and occupied 45% (34% in moderate; 11% in strong) of the Red+ area. This MHW event lasted for 39 days (5 days as a strong MHW and 31 days as a moderate MHW) with a mean intensity of 0.92 °C (equivalent to accumulative intensity of 36.03 °C-day), as seen in Figure 10. At a 95% level of significance, the MHW intensity during this study case showed a significant correlation with SLHF (R = −0.52, n = 39), as seen in Figure 11a. In the same context, the other studied surface radiation components (SSHF, SNTR and SNSR) had an insignificant correlation with the marine heat wave intensity. In the same context, F_{loss} shows a significant correlation with the MHW intensity (R = −0.55, n = 39), as seen in Figure 11e.

The effect of this MHW event on the chlor_a concentration is not discussed due to the missing chlor_a concentration data during this study case.

The second study case (peaked on 18 October 2010) centered around the eastern side of the Gulf of Aden (12.875° N and 50.875° E), occupied 37% of the Red+ (36% in moderate; 1% in strong) area and lasted for 25 days (4 days as a strong MHW and 16 days as a moderate MHW), with a mean intensity of 2.02 °C (equivalent to accumulative intensity of 50.5 °C-day), as seen in Figure 12. There is a significant correlation between the marine heat wave intensity during this study case and different components of the surface radiation at the 95% significance level, most markedly with SLHF (R = 0.65, n = 25), as seen in Figure 13. Similarly, F_{loss} shows a significant correlation with the MHW intensity (R = 0.64, n = 25), as seen in Figure 13e.

During this second study case, the chlor_a concentration is significantly lower than its climatological values over 82% of the time (Figure 14a), partly showing the effect of MHW on the chlor_a concentration and hence the Red Sea ecology. Moreover, there is a significant correlation between the MHW intensity and chlor_a concentration during this study case (R = −0.40, n = 25) at the 95% significance level, as seen in Figure 14b.

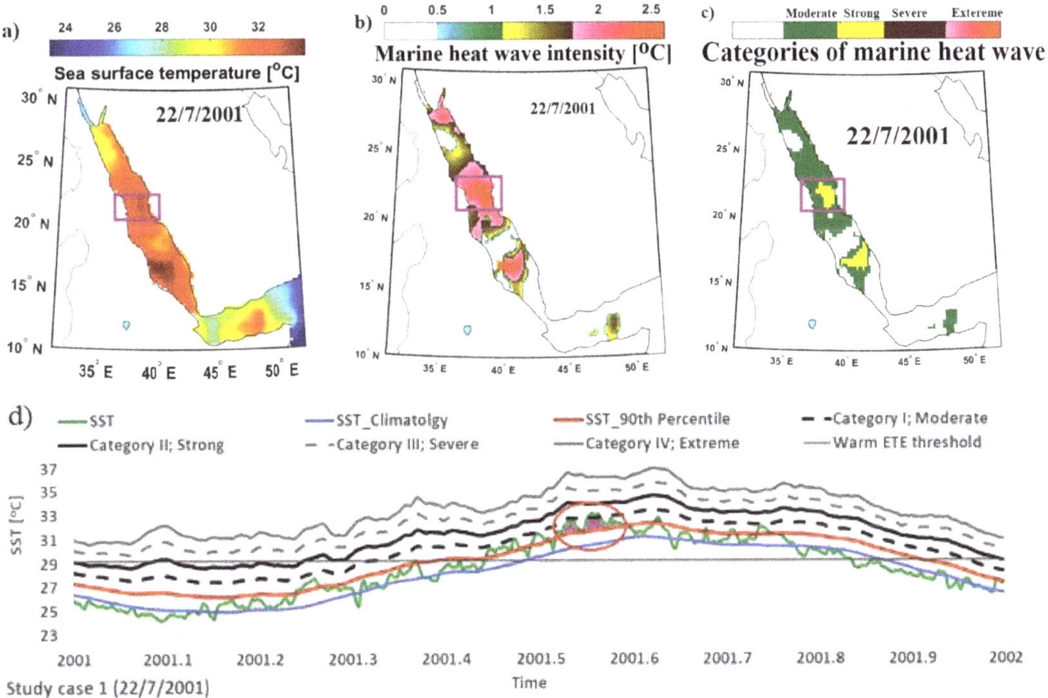

Figure 10. Representative of the MHW event (study case 1) showing the sea surface temperature (**a**), intensity [°C] of marine heat wave event (**b**) and category of that event (**c**) at the peak of the event and the time series during the event's year (**d**), spatially averaged over an area around the maximum intensity grid, which is shown in bubbles.

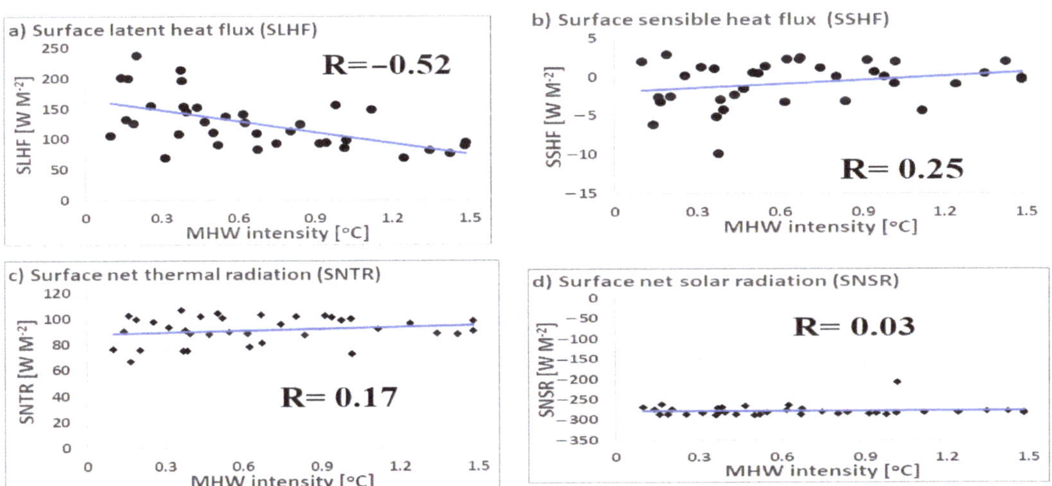

Figure 11. *Cont.*

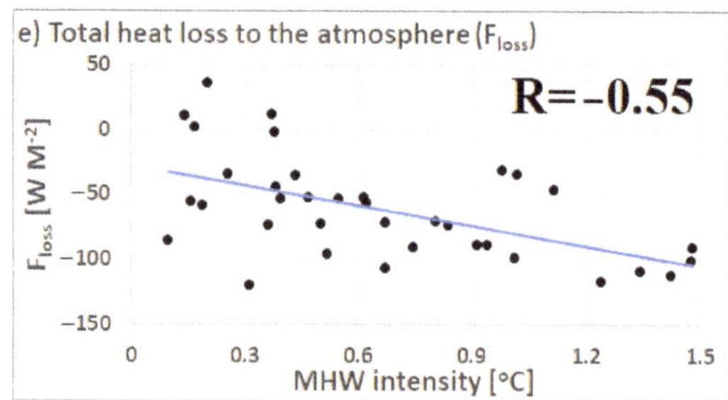

Figure 11. Scatter plots of marine heat wave intensity during the study case 1 (10 July 2001 to 22 August 2001) and surface latent heat flux (**a**), surface sensible heat flux (**b**), surface net thermal radiation (**c**), surface net solar radiation (**d**), and total heat loss to the atmosphere (**e**).

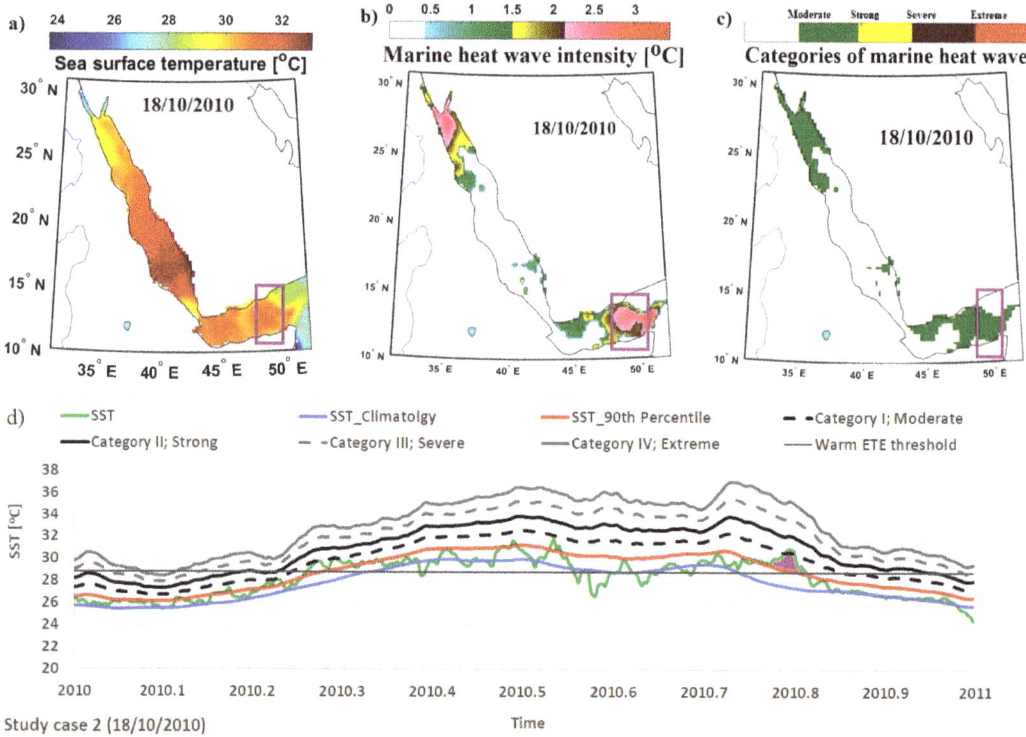

Figure 12. Representative of the MHW event (study case 2) showing the sea surface temperature (**a**), intensity [°C] of marine heat wave event (**b**) and category of that event (**c**) at the peak of the event and the time series during the event's year (**d**), spatially averaged over an area around the maximum intensity grid, which is shown in bubbles.

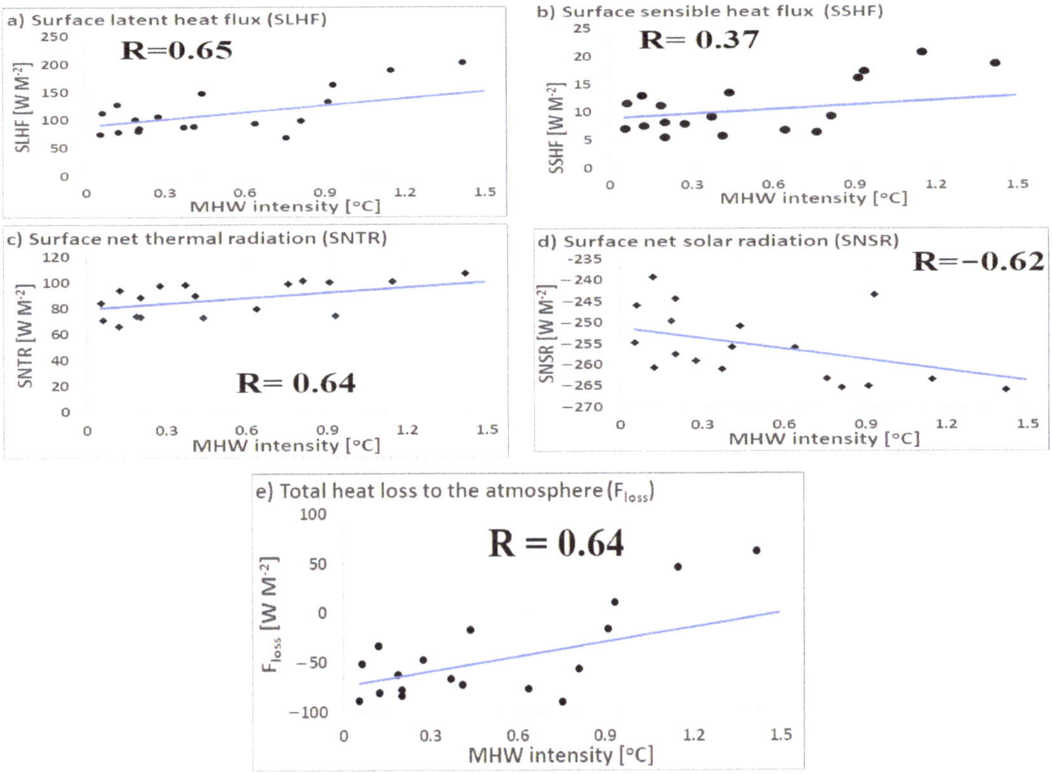

Figure 13. Scatter plots of marine heat wave intensity during the study case 2 (29 September 2010 to 23 October 2010) and surface latent heat flux (**a**), surface sensible heat flux (**b**), surface net thermal radiation (**c**), surface net solar radiation (**d**), and total heat loss to the atmosphere (**e**).

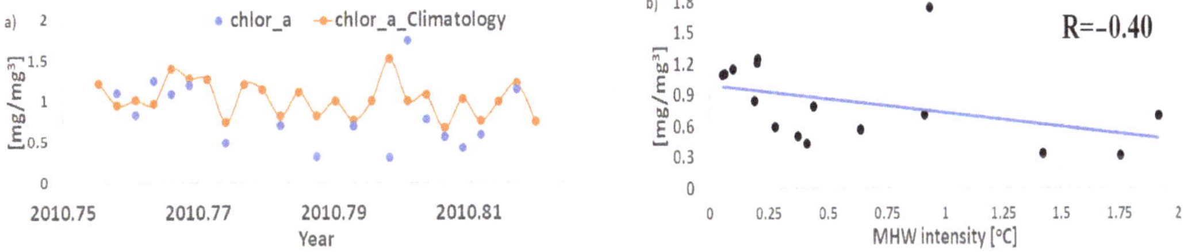

Figure 14. Representative of the chlorophyll_concentration_in_sea_water (chlor_a) values and its climatological values during the study case 2, which extends from 29 September 2010 to 23 October 2010 (**a**); scatter plots of marine heat wave intensity and chlor_a concentration during the study case 2 (**b**).

The third study case, which peaked on 18th of July 2019, extended for 24.1% (23% in moderate; 0.5% in strong; 0.4% in severe and 0.2% in extreme) of the study area, most markedly over the strait of Bab al Mandab (Figure 15). Over this strait, this MHW had a mean intensity of 2.99 °C over a duration of 26 days (equivalent to accumulative intensity of 77.84 °C-day): 3 days as an extreme MHW, 5 days as a severe MHW, 12 days as a strong MHW and 6 days as a moderate MHW. At the 95% level of significance, the marine heat wave intensity during this study case showed a significant correlation only with SNTR ($R = -0.45$, $n = 26$), as seen in Figure 16c. In the same context, the other studied surface radiation components (SLHF, SSHF, SNSR and F_{loss}) had an insignificant correlation with the marine heat wave intensity.

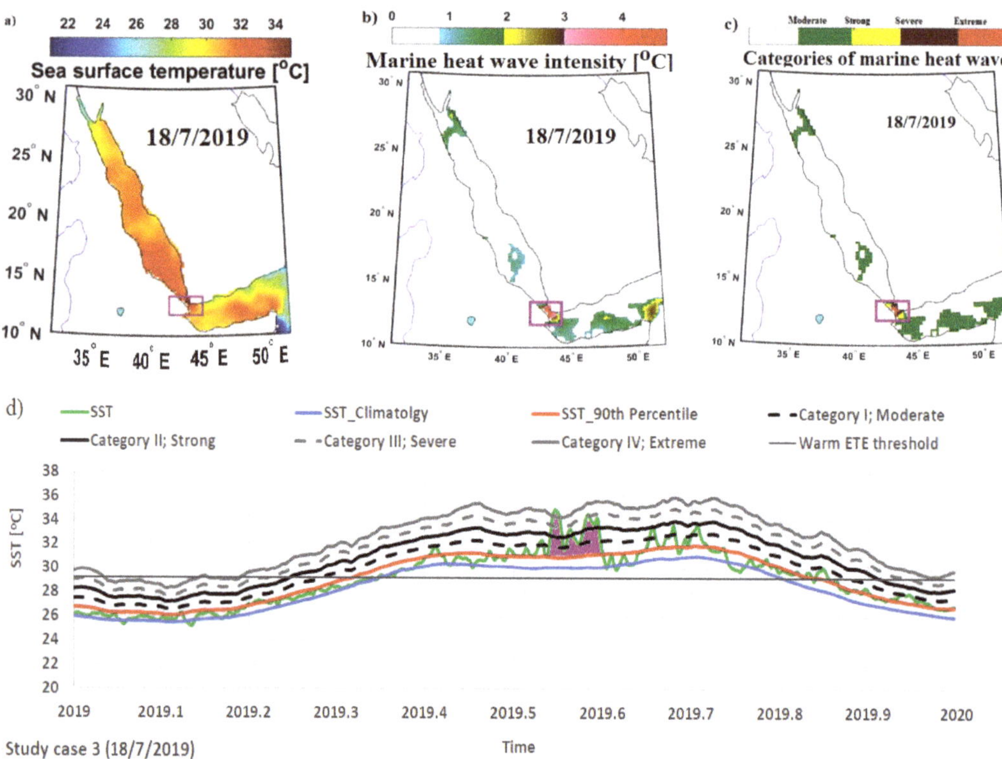

Figure 15. Representative of the MHW event (study case 3) showing the sea surface temperature (**a**), intensity [°C] of marine heat wave event (**b**) and category of that event (**c**) at the peak of the event and the time series during the event's year (**d**), spatially averaged over an area around the maximum intensity grid, which is shown in bubbles.

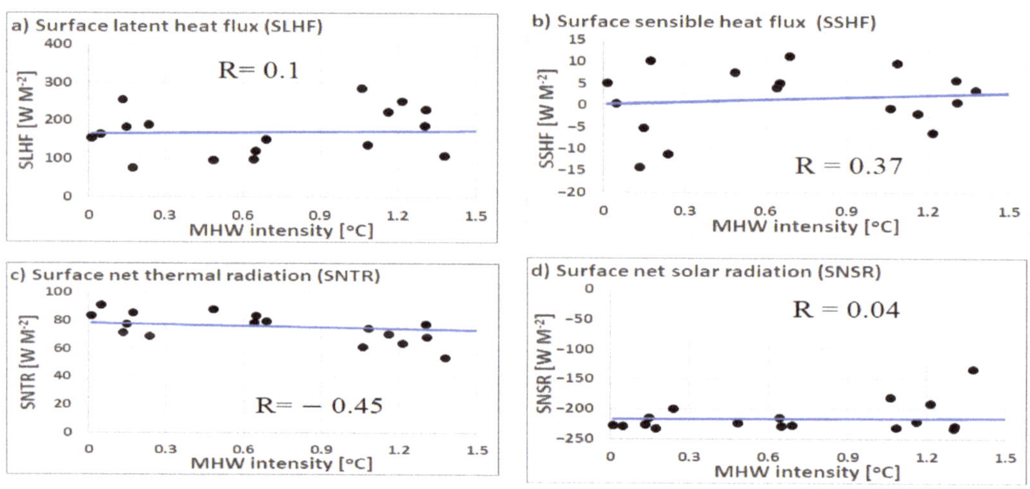

Figure 16. *Cont.*

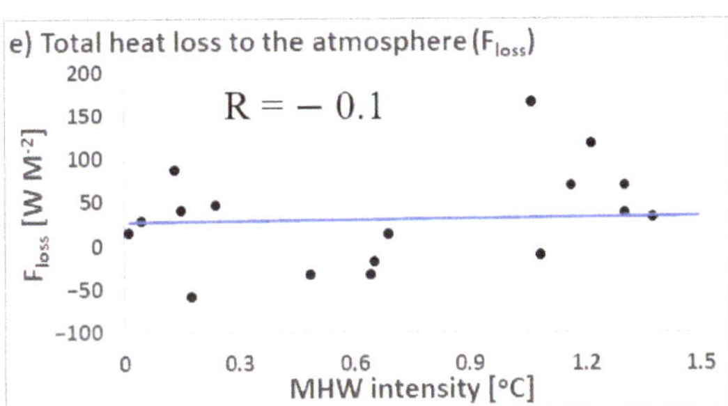

Figure 16. Scatter plots of marine heat wave intensity during the study case 3 (27 June 2019 to 22 July 2019) and surface latent heat flux (**a**), surface sensible heat flux (**b**), surface net thermal radiation (**c**), surface net solar radiation (**d**), and total heat loss to the atmosphere (**e**).

During this study case, the link between this MHW event and the chlor_a concentration is not discussed due to the missing chlor_a concentration data.

The fourth study case, which peaked on 13 September 2020, extended over 85% (62% in moderate; 23% in strong) of the study area, most markedly around the northern side of the Red Sea (27.625° N and 34.875° E), as seen in Figure 17. Over the area centered around (27.625° N and 34.875° E), the MHW duration extended for a 98-day period from 30 August 2020 to 5 December 2020. This MHW had a mean intensity of 2.56 °C over a 98-day duration (equivalent to an accumulative intensity of 251.3 °C- day): 42 days as a strong MHW and 54 days as a moderate MHW. At the 95% significance level, the correlation between the marine heat wave intensity during this study case and different components of surface radiation, together with the F_{loss}, are significant, most markedly with SNTR (R = −0.65, *n* = 98), as seen in Figure 18.

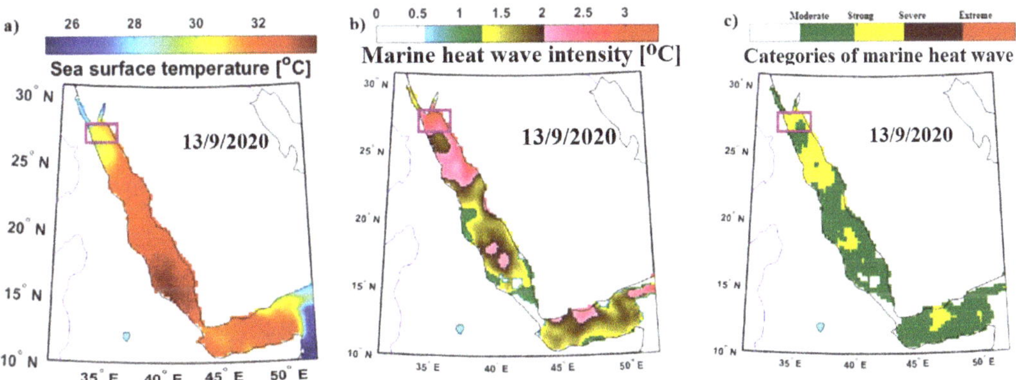

Figure 17. *Cont.*

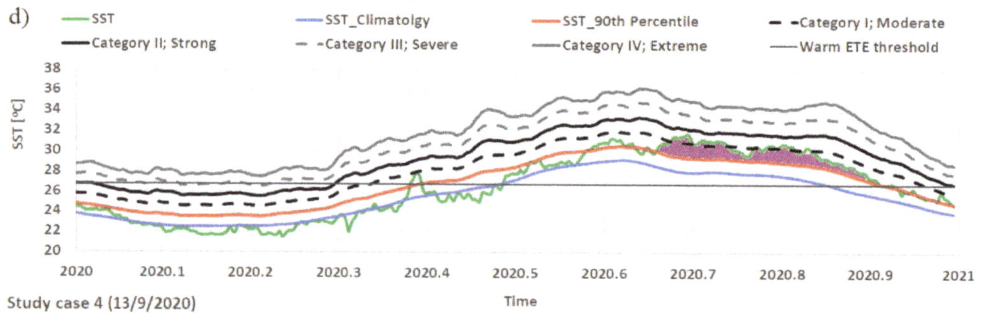

Figure 17. Representative of the MHW event (study case 4) showing the sea surface temperature (**a**), intensity [°C] of marine heat wave event (**b**) and category of that event (**c**) at the peak of the event and the time series during the event's year (**d**), spatially averaged over an area around the maximum intensity grid, which is shown in bubbles.

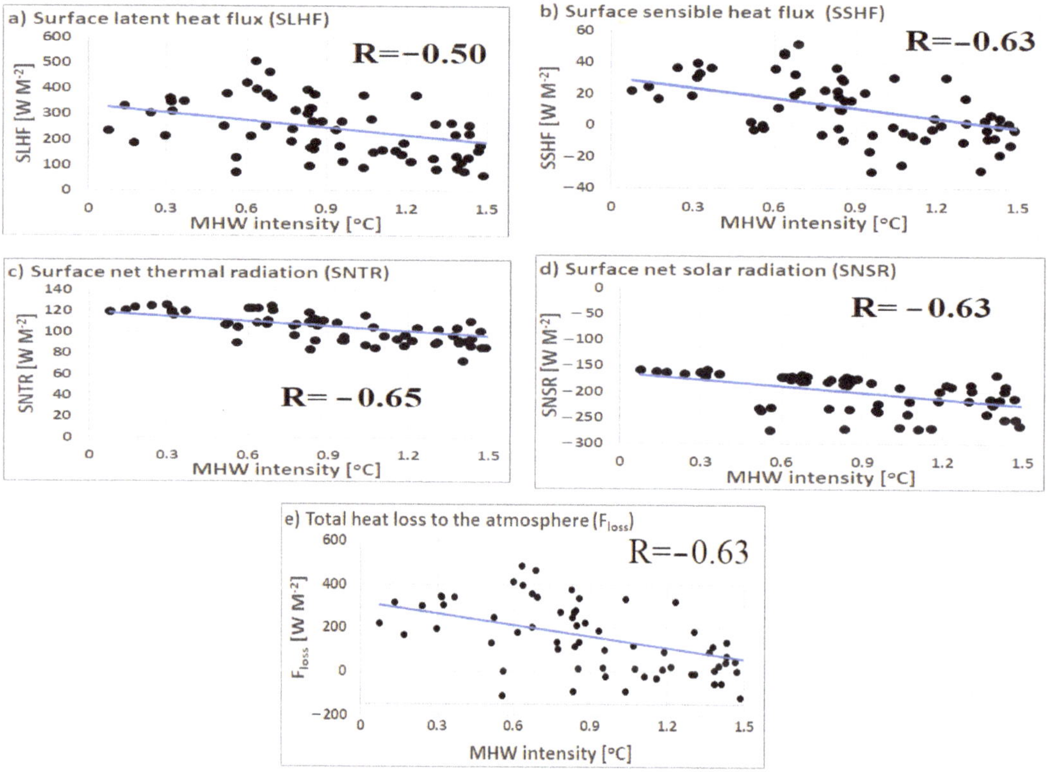

Figure 18. Scatter plots of marine heat wave intensity during the study case 4 (30 August 2020 to 5 December 2020) and surface latent heat flux (**a**), surface sensible heat flux (**b**), surface net thermal radiation (**c**), surface net solar radiation (**d**), and total heat loss to the atmosphere (**e**).

During this study case, the chlor_a concentration is significantly lower than its climatological values over 100% of the time (Figure 19a), confirming the effect of MHWs on the Red Sea ecology similarly to study case 2. In the same context and at the 95% significance level, the MHW intensity and chlor_a concentration had a significant correlation (R = −0.43, *n* = 98) during this study case, as seen in Figure 19b.

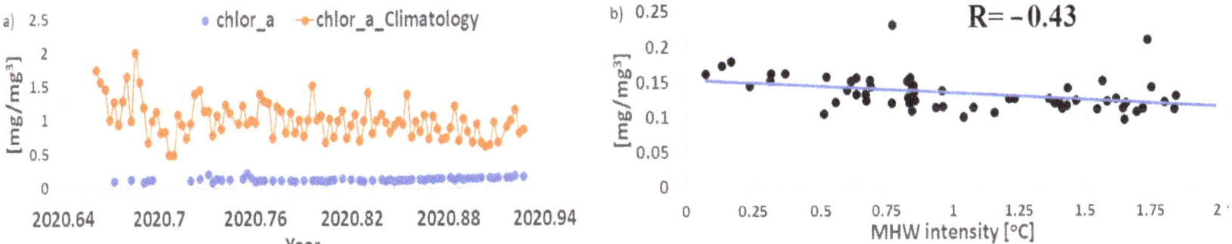

Figure 19. Representative of the chlorophyll_concentration_in_sea_water (chlor_a) values and its climatological values during the study case 4, which extends from 30 August 2020 to 5 December 2020 (**a**); scatter plots of marine heat wave intensity and chlor_a concentration during the study case 2 (**b**).

3.4. Main Spatiotemporal Characteristics of Marine MHWs over the Red+

The annual average of the MHWs intensity over Red+ has significant spatial variability, where the lowest values (<1.2 °C) were found along the Saudi Arabia coast at around 18.875° N latitude, and the highest values (>2 °C) were found to the south of the Gulf of Aqaba at around 29.375° N latitude, as described by Figure 20a. On the other hand, the MHWs duration exhibits a markedly spatial pattern and reached its maximum values along the Saudi Arabia coast at around 28.375° N latitude and its minimum values in the southern part of the Red Sea at around 15.625° N, as seen in Figure 20b. Moreover, the MHW frequency shows a different spatial pattern, where its maximum values were found along the Saudi Arabia coast near 19.375° N and its minimum values at the north part of the Gulf of Suez near 29.625° N, as seen in Figure 20c. In general, the Red Sea displays a meridional gradient of an increasing annual average of the MHW intensity and MHW duration from north to south. Conversely, the Red Sea displays a meridional gradient of a decreasing annual average of the MHW frequency from north to south.

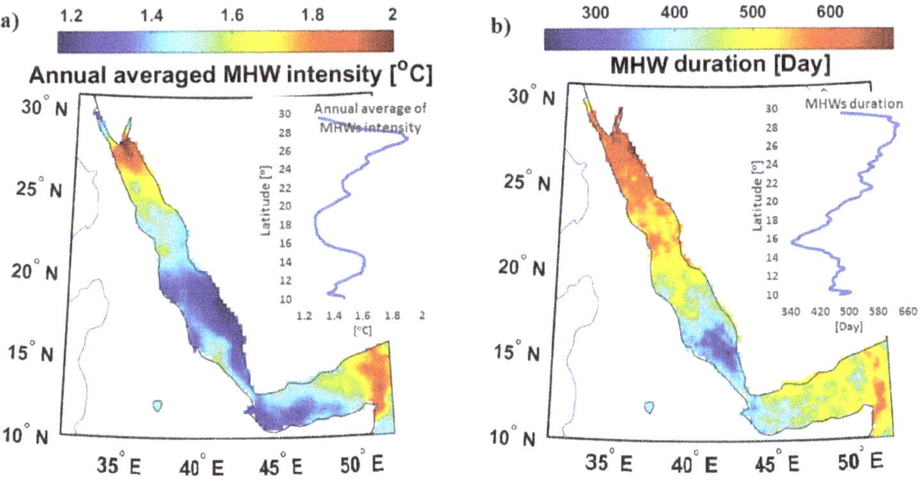

Figure 20. *Cont.*

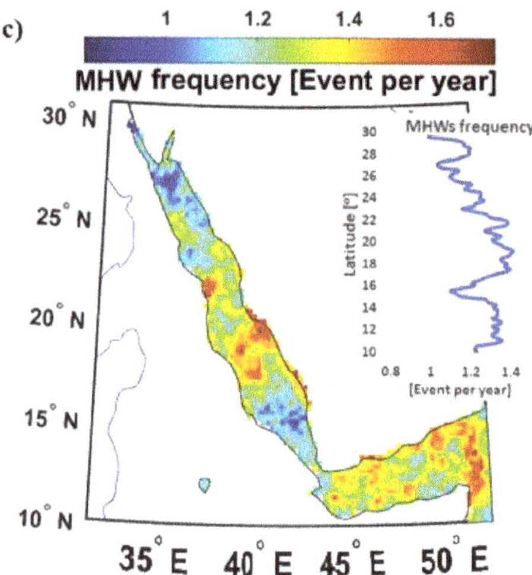

Figure 20. Marine heat wave (MHW) spatial/temporal characteristics over Red+. The latitudinal changes in annual average MHW intensity (**a**), MHW duration (**b**), and MHW frequency (**c**) are also shown.

4. Summary Discussion

Extreme Red+ sea surface temperature is calculated based on analyses of seasonal mean time series from 1982 to 2020, confirming that the thresholds of warm (cold) ETEs is 30.03 (25.13) °C using a 90th (10th) percentile definition. Warm ETEs will be used as an additional thermal restriction to MHWs' threshold values.

The changeable (based on each day of the year) climatological SST mean and 90th percentile threshold was used to identify the MHW events. To avoid the cold season period, warm ETEs were used as an additional thermal restriction to the MHWs' threshold values.

From 1982 to 2020, there are 28 different MHW events that extend for 378 days; 360 of those days are in the moderate category, whereas the other 18 days are in a strong category. On the other hand, the MHW average intensity, annual duration and frequency described a significant increasing trend over Red+ from 1982 to 2020. These results follow the global trend pattern (Oliver et al. [3]). The longest MHWs over Red+ extended for 57 days from 31 August 2020 to 26 October 2020, with a maximum (mean) intensity of 1.76 (1.11) °C and a cumulative intensity of 63.78 °C-day. For the longest MHW event, the currently identified MHW duration is approximately 3.2 times more than previously detected by Shaltout [6]. This difference in determining the longest duration of MHWs is due to the fact that the current study extends into 2020 and also because different methods were used. In general, the MHW over Red+ is in a moderate category, with a frequency of 1.38 events annually. The average value of an event intensity and duration is 0.8 °C and 22.5 days, respectively. Furthermore, the annual average days of MHW over Red+ is 9 days, which agrees with the previous finding of Shaltout [6].

During El Niño periods, the MHW intensity tends to reach its maximum value. In the same context, the ISM rarely explains the MHW intensity (the higher values of the MHWs' intensity may occur simultaneously with lower ISMI values and, at another time, may occur with higher ISMI values). Furthermore, the sea surface radiation budget of different components may describe critical conditions of the occurrence of MHW events (more than 7 W m^{-2} of F_{loss} going to the ocean). As ENSO and ISMI indexes, together with different components of the sea surface radiation budget, cannot completely describe the controls of MHWs, there is another climatic mechanism that, if coupled with them, will describe the MHWs' characteristics. This mechanism merits our consideration and will be discussed in our future work by describing a new climatic index for the study area.

There is a markedly interannual variability for the studied MHW property over Red+, which is followed well by the global properties (Oliver et al. [35]). The pattern of the mean MHW frequency, duration, intensity and SST were positively correlated over Red+, indicating that global warming is the main reason for the positive trends of the MHW frequency, duration and intensity. On a global scale, the average duration and frequency of MHWs were negatively correlated (Oliver et al. [35]), indicating that their relations over Red+ (depending on the current result) do not follow the global patterns. Moreover, higher frequency and intensity values on a global scale appear to relate to El Niño period events (Oliver et al., 2018); however, over Red+, only the intensity is clearly related to El Niño periods. On the other side, the MHWs' intensity has a clearer relation with ISMI than the frequency and duration have with ISMI. This encourages us to find and describe a new valid climatic indicator for the study area in our future study.

For further details about the MHW characteristics, the longest four MHW events were analyzed in depth. These further analyses prove that the chlor_a concentration has lower values than its climatic values during the MHW events, providing early awareness about the impact of MHWs as a heat stress on various marine sectors while shedding light on decision makers in finding a suitable regional climate policy to cope with global warming issues.

Finally, the spatiotemporal analysis of MHWs confirmed that the spatial distribution of the MHW annual average intensity, duration and frequency had a distinct spatial pattern, which agrees with the previous finding of Genevier et al. [20]. The northern region of Red+ (south of the Gulf of Suez and Aqaba) witnessed the most intense MHW events, which disagrees with the previous finding of Genevier et al. [20]. On the other hand, the eastern coast of the northern region had the greatest number of MHW days, which agrees with the previous finding of Genevier et al. [20]. The northeastern coast of the southern region had the most frequent events, which agrees with Genevier et al. [20]. Generally, the current results agree with the previous finding of Genevier et al. [20] concerning the MHW intensity and duration spatial distribution. The only exception in the spatial pattern of the MHW duration is possibly due to the use of warm ETEs as an additional thermal restriction to MHWs' threshold values.

Author Contributions: Conceptualization, M.S. and K.T.; methodology, M.S.; software, M.S. and A.B.; validation, M.S., N.G., N.B. and K.T.; formal analysis, N.G.; investigation, N.B.; resources, A.B.; data curation, M.S. and K.T.; writing original draft preparation, M.S.; writing review and editing, M.S. and K.T.; visualization, M.S. and N.G.; supervision, M.S.; project administration, A.B.; funding acquisition, A.B. All authors have read and agreed to the published version of the manuscript.

Funding: This research was funded by the Deputyship for Research & Innovation, Ministry of Education in Saudi Arabia, grant number 1213 and "The APC was funded by the Deputyship for Research & Innovation, Ministry of Education in Saudi Arabia".

Institutional Review Board Statement: Not applicable.

Informed Consent Statement: Not applicable.

Data Availability Statement: Not applicable.

Acknowledgments: The authors extend their appreciation to the Deputyship for Research & Innovation, Ministry of Education in Saudi Arabia for funding this research work through the project number 1213.

Conflicts of Interest: The authors declare no conflict of interest.

References

1. Hobday, A.J.; Alexander, L.V.; Perkins-Kirkpatrick, S.; Smale, D.; Straub, S.; Oliver, E.; Benthuysen, J.; Burrows, M.; Donat, M.; Feng, M.; et al. A hierarchical approach to defining marine heatwaves. *Prog. Oceanogr.* **2016**, *141*, 227–238. [CrossRef]
2. Sparnocchia, S.; Schiano, M.E.; Picco, P.; Bozzano, R.; Cappelletti, A. The anomalous warming of summer 2003 in the surface layer of the Central Ligurian Sea (Western Mediterranean). *Ann. Geophys.* **2006**, *24*, 443–452. [CrossRef]

3. Oliver, E.C.J.; Benthuysen, J.; Bindoff, N.; Hobday, A.J.; Holbrook, N.J.; Mundy, C.; Perkins-Kirkpatrick, S. The unprecedented 2015/16 Tasman Sea marine heatwave. *Nat. Commun.* **2017**, *8*, 16101. [CrossRef]
4. Holbrook, N.J.; Scannell, H.A.; Gupta, A.S.; Benthuysen, J.A.; Feng, M.; Oliver, E.C.J.; Alexander, L.V.; Burrows, M.T.; Donat, M.G.; Hobday, A.J.; et al. A global assessment of marine heatwaves and their drivers. *Nat. Commun.* **2019**, *10*, 1–13. [CrossRef]
5. Chaidez, V.; Dreano, D.; Agusti, S.; Duarte, C.M.; Hoteit, I. Decadal trends in Red Sea maximum surface temperature. *Sci. Rep.* **2017**, *7*, 8144. [CrossRef]
6. Shaltout, M. Recent sea surface temperature trends and future scenarios for the Red Sea. *Oceanologia* **2019**, *61*, 484–504. [CrossRef]
7. Olita, A.; Sorgente, R.; Ribotti, A.; Natale, S.; Gaberšek, S. Effects of the 2003 European heatwave on the Central Mediterranean Sea surface layer: A numerical simulation. *Ocean Sci. Discuss.* **2006**, *3*, 85–125.
8. Ibrahim, O.; Mohamed, B.; Nagy, H. Spatial Variability and Trends of Marine Heat Waves in the Eastern Mediterranean Sea over 39 Years. *J. Mar. Sci. Eng.* **2021**, *9*, 643. [CrossRef]
9. Pearce, A.F.; Feng, M. The rise and fall of the 'marine heat wave' off Western Australia during the summer of 2010/11. *J. Mar. Syst.* **2013**, *112*, 139–156. [CrossRef]
10. Benthuysen, J.A.; Oliver EC, J.; Feng, M.; Marshall, A.G. Extreme marine warming across tropical Australia during austral summer 2015–16. *J. Geophys. Res.* **2018**, *123*, 1301–1326. [CrossRef]
11. Bond, N.A.; Cronin, M.F.; Freeland, H.; Mantua, N. Causes and impacts of the 2014 warm anomaly in the NE Pacific. *Geophys. Res. Lett.* **2015**, *42*, 3414–3420. [CrossRef]
12. Chen, K.; Gawarkiewicz, G.G.; Lentz, S.J.; Bane, J.M. Diagnosing the warming of the Northeastern US Coastal Ocean in 2012: A linkage between the atmospheric jet stream variability and ocean response. *J. Geophys. Res. Ocean.* **2014**, *119*, 218–227. [CrossRef]
13. Hughes, T.P.; Kerry, J.T.; Álvarez-Noriega, M.; Álvarez-Romero, J.; Anderson, K.; Baird, A.H.; Babcock, R.; Beger, M.; Bellwood, D.R.; Berkelmans, R.; et al. Global warming and recurrent mass bleaching of corals. *Nature* **2017**, *543*, 373–377. [CrossRef] [PubMed]
14. Caputi, N.; Kangas, M.; Denham, A.; Feng, M.; Pearce, A.; Hetzel, Y.; Chandrapavan, A. Management adaptation of invertebrate fisheries to an extreme marine heat wave event at a global warming hot spot. *Ecol. Evol.* **2016**, *6*, 3583–3593. [CrossRef]
15. Garrabou, J.; Coma, R.; Bensoussan, N.; Bally, M.; Chevaldonné, P.; Ciglianos, M.; Díaz, D.; Harmelin, J.G.; Gambi, M.C.; Kersting, D.K.; et al. Mass mortality in Northwestern Mediterranean rocky benthic communities: Effects of the 2003 heat wave. *Glob. Chang. Biol.* **2009**, *15*, 1090–1103. [CrossRef]
16. Mills, K.E.; Pershing, A.J.; Brown, C.J.; Chen, Y.; Chiang, F.-S.; Holland, D.; Lehuta, S.; Nye, J.; Sun, J.C.; Thomas, A.C.; et al. Fisheries Management in a Changing Climate: Lessons From the 2012 Ocean Heat Wave in the Northwest Atlantic. *Oceanography* **2013**, *26*. [CrossRef]
17. Cavole, L.M.; Demko, A.M.; Diner, R.E.; Giddings, A.; Koester, I.; Pagniello, C.M.L.S.; Paulsen, M.-L.; Ramirez-Valdez, A.; Schwenck, S.M.; Yen, N.K. Biological impacts of the 2013–2015 warm-water anomaly in the Northeast Pacific: Winners, losers, and the future. *Oceanography* **2016**, *29*, 273–285. [CrossRef]
18. Bindoff, N.L.; Stott, P.A.; AchutaRao, K.M.; Allen, M.R.; Gillett, N.; Gutzler, D.; Hansingo, K.; Hegerl, G.; Hu, Y.; Jain, S.; et al. Detection and attribution of climate change: From global to regional. In *Climate Change 2013: The Physical Science Basis*; Contribution of Working Group 1 to the Fifth Assessment Report of the Intergovernmental Panel on Climate Change; Stocker, B.D., Qin, D., Plattner, G.-K., Tignor, M.M.B., Allen, S.K., Boschung, J., Nauels, A., Xia, Y., Bex, V., Midgley, P.M., Eds.; Cambridge University Press: Cambridge, UK, 2013; pp. 867–952.
19. Hodgkinson, J.H.; Hobday, A.J.; Pinkard, E.A. Climate adaptation in Australia's resource-extraction industries: Ready or not? *Reg. Environ. Chang.* **2014**, *14*, 1663–1678. [CrossRef]
20. Genevier, L.G.C.; Jamil, T.; Raitsos, D.E.; Krokos, G.; Hoteit, I. Marine heatwaves reveal coral reef zones susceptible to bleaching in the Red Sea. *Glob. Chang. Biol.* **2019**, *25*, 2338–2351. [CrossRef]
21. Darmaraki, S.; Somot, S.; Sevault, F.; Nabat, P.; Narvaez, W.D.C.; Cavicchia, L.; Djurdjevic, V.; Li, L.; Sannino, G.; Sein, D.V. Future evolution of Marine Heatwaves in the Mediterranean Sea. *Clim. Dyn.* **2019**, *53*, 1371–1392. [CrossRef]
22. Reynolds, R.W.; Smith, T.M.; Liu, C.; Chelton, D.B.; Casey, K.; Schlax, M.G. Daily High-Resolution-Blended Analyses for Sea Surface Temperature. *J. Clim.* **2007**, *20*, 5473–5496. [CrossRef]
23. Reynolds, R.W. What's New in Version 2 of Daily Optimum Interpolation (OI) Sea Surface Temperature (SST) Analysis. 2009. 10p. Available online: https://www.ncdc.noaa.gov/sites/default/files/attachments/Reynolds2009_oisst_daily_v02r00_version2 -features.pdf (accessed on 19 August 2021).
24. Banzon, V.; Reynolds, R.; National Center for Atmospheric Research Staff (Eds.) *The Climate Data Guide: SST Data: NOAA Optimal Interpolation (OI) SST Analysis, Version 2 (OISSTv2) 1x1*. 2018. Available online: https://climatedataguide.ucar.edu/climate-data/ sst-data-noaa-optimal-interpolation-oi-sst-analysis-version-2-oisstv2-1x1 (accessed on 17 December 2018).
25. Worley, S.J.; Woodruff, S.D.; Reynolds, R.W.; Lubker, S.J.; Lott, N. ICOADS release 2.1 data and products. *Int. J. Climatol.* **2005**, *25*, 823–842. [CrossRef]
26. Karnauskas, K.B.; Jones, B.H. The Interannual Variability of Sea Surface Temperature in the Red Sea from 35 Years of Satellite and In Situ Observations. *J. Geophys. Res. Oceans* **2018**, *123*, 5824–5841. [CrossRef]
27. SeaWiFS Mission Page. NASA Ocean Biology Processing Group, Greenbelt, MD, USA. Maintained by NASA Ocean Biology Distibuted Active Archive Center (OB.DAAC), Goddard Space Flight Center, Greenbelt MD. 2019. Available online: https: //oceancolor.gsfc.nasa.gov/data/10.5067/ORBVIEW-2/SEAWIFS/L2/OC/2018/ (accessed on 16 August 2021).

28. Brewin, R.J.; Raitsos, D.E.; Pradhan, Y.; Hoteit, I. Comparison of chlorophyll in the Red Sea derived from MODIS-Aqua and in vivo fluorescence. *Remote. Sens. Environ.* **2013**, *136*, 218–224. [CrossRef]
29. Eladawy, A.; Nadaoka, K.; Negm, A.; Abdel-Fattah, S.; Hanafy, M.; Shaltout, M. Characterization of the northern Red Sea's oceanic features with remote sensing data and outputs from a global circulation model. *Oceanologia* **2017**, *59*, 213–237. [CrossRef]
30. Zhang, T.; Hoell, A.; Perlwitz, J.; Eischeid, J.; Murray, D.; Hoerling, M.; Hamill, T.M. Towards Probabilistic Multivariate ENSO Monitoring. *Geophys. Res. Lett.* **2019**, *46*, 10532–10540. [CrossRef]
31. Wang, B.; Wu, R.; Lau, K. Interannual variability of Asian summer monsoon: Contrast between the Indian and western North Pacific-East Asian monsoons. *J. Climatol.* **2001**, *14*, 4073–4090. [CrossRef]
32. Hersbach, H.; Bell, B.; Berrisford, P.; Hirahara, S.; Horanyi, A.; Muñoz-Sabater, J.; Nicolas, J.; Peubey, C.; Radu, R.; Schepers, D.; et al. The ERA5 global reanalysis. *Q. J. R. Meteorol. Soc.* **2020**, *146*, 1999–2049. [CrossRef]
33. Copernicus Climate Change Service (C3S). ERA5: Fifth Generation of ECMWF Atmospheric Reanalyses of the Global Climate. Copernicus Climate Change Service Climate Data Store (CDS). 2017. Available online: https://cds.climate.copernicus.eu/cdsapp#!/home (accessed on 29 January 2020).
34. Hobday, A.; Oliver, E.C.J.; Gupta, A.S.; Benthuysen, J.A.; Burrows, M.T.; Donat, M.G.; Holbrook, N.J.; Moore, P.J.; Thomsen, M.S.; Wernberg, T.; et al. Categorizing and naming marine heatwaves. *Oceanography* **2018**, *31*, 162–173.
35. Oliver, E.C.J.; Donat, M.G.; Burrows, M.; Moore, P.; Smale, D.A.; Alexander, L.V.; Benthuysen, J.; Feng, M.; Gupta, A.S.; Hobday, A.J.; et al. Longer and more frequent marine heatwaves over the past century. *Nat. Commun.* **2018**, *9*, 1324. [CrossRef]

Journal of
*Marine Science
and Engineering*

Spatiotemporal Variability and Trends of Marine Heat Waves in the Red Sea over 38 Years

Bayoumy Mohamed [1,2], Hazem Nagy [1,3,*] and Omneya Ibrahim [1]

[1] Oceanography Department, Faculty of Science, Alexandria University, Alexandria 21500, Egypt;
mohamedb@unis.no (B.M.); omneya.ibrahim@alexu.edu.eg (O.I.)

[2] Department of Arctic Geophysics, University Centre in Svalbard, 9171 Longyearbyen, Norway

[3] Marine Institute, Oranmore, Co Galway, Ireland

* Correspondence: hazem.nagy@marine.ie; Tel.: +353-(89)-985-494

Abstract: Marine heat waves (MHWs) can have catastrophic consequences for the socio-environmental system. Especially in the Red Sea, which has the world's second longest coral reef system. Here, we investigate the sea surface temperature (SST) variability and trends, as well as the spatiotemporal characteristics of marine heat waves (MHWs) in the Red Sea, using high resolution daily gridded (1/20°) SST data obtained from the Copernicus Marine Environment Monitoring Service (CMEMS) for the period 1982–2019. Results show that the average warming rate was about $0.342 \pm 0.047\ °C/$decade over the entire Red Sea over the whole study period. The Empirical Orthogonal Function (EOF) analysis reveals that the maximum variability is over the central part of the Red Sea, while the minimum variability is in the southernmost part of the Red Sea. Over the last two decades (2000–2019), we have discovered that the average MHW frequency and duration increased by 35% and 67%, respectively. The results illustrate that the MHW frequency and duration trends have increased by 1.17 counts/decade and 1.79 days/decade, respectively, over the study period. The highest annual MHW frequencies were detected in the years 2018, 2019, 2010, and 2017. A strong correlation (R = 0.89) was found between the annual MHW frequency and the annual mean SST.

Keywords: Red Sea; marine heat wave; sea surface temperature; duration; frequency

Citation: Mohamed, B.; Nagy, H.; Ibrahim, O. Spatiotemporal Variability and Trends of Marine Heat Waves in the Red Sea over 38 Years. *J. Mar. Sci. Eng.* **2021**, *9*, 842. https://doi.org/10.3390/jmse9080842

Academic Editor: Francisco Pastor Guzman

Received: 15 July 2021
Accepted: 1 August 2021
Published: 4 August 2021

Publisher's Note: MDPI stays neutral with regard to jurisdictional claims in published maps and institutional affiliations.

1. Introduction

Heat waves are characterized as extended periods of unusually hot weather that have become more frequent and longer in recent decades, posing a threat to human health and ecosystems. A similar phenomenon known as a 'Marine Heat Wave' (MHW) has been observed in the oceans, endangering marine ecosystems and productivity [1–4]. The most recent definition of MHWs is discrete periods of abnormally warm sea surface temperatures that can last from five days to months above the 90th percentile threshold of SST climatology [5–7]. Despite extensive knowledge of global SST changes, the study and detection of MHWs and associated climate processes remains limited [2].

The Red Sea has a rich and diverse ecosystem [8]. It has the world's second longest coral reef system [9]. The importance of the Red Sea is highlighted by its high tourism volume and aquaculture, and also its relation to the gas and oil and fishing industries. Furthermore, the Red Sea is a hub for shipping activity, connecting European ports with China and Eastern Asia through Suez Canal.

The Red Sea has a negative water balance, which means that evaporation exceeds precipitation and combined river runoff [10,11]. It is considered to be an arid climate region with little precipitation and runoff, as there is a lack of any permanent rivers that flow into it [12–14].

The impact and consequences of MHW on Red Sea marine life are now critical for decision makers and stakeholders. There are no published MHW studies of the Red Sea. However, some studies in the Rea Sea have discovered a link between extreme heat events

125

such as ocean heat waves and the spread of the El Nino-Southern Oscillation (ENSO) [15]. In addition, the authors have found several SST studies for the Red Sea region, addressing trends and changes in ocean surface temperature [16–27].

According to [21], the annual mean SST has changed from 27.4 °C from 1985 to 1993 to 28.1 °C from 1994 to 2007, which was accompanied with significant warming in the mid-1990s. The authors of [15] discovered that the Red Sea maximum SST is warming at a rate of 0.17 °C/decade, with the northern Red Sea alone warming from 0.4 to 0.45 °C/decade from 1982 to 2015, which exceeds the global trend of 0.11 °C/decade [15]. The amplification of recent summer warming and decadal variability of the Red Sea SST was studied in [25]. They found that the significant warming which began in the mid-1990s precisely corresponded with the onset of a positive phase of the Atlantic Multi-Decadal Oscillation (AMO) and a negative phase of the Silk Road Pattern (SRP) in 1997.

In this study, we have investigated the spatiotemporal characteristics and occurrence of MHWs in the Red Sea from 1982 to 2019 using high resolution SST OSTIA (Operational SST and Ice Analysis) data (0.05° × 0.05°). In addition, the SST spatial trend and variability were examined.

Section 2 describes the data and methodology of SST analysis. Section 3 reports the SST climatology, trend, and interannual variability. Section 4 shows the MHW results over the study period. Section 5 presents the discussion and Section 6 provides the conclusions.

2. Data and Methods

The Red Sea is an almost enclosed marginal tropical sea characterized by an extended shape that separates the African and Asian continents and is connected to the Indian Ocean through the Bab El-Mandab Strait, as shown in Figure 1. It has a moderate size of 4.5×10^6 km^3, which is roughly one fifth the size of the Mediterranean Sea. The Bab El-Mandeb strait, which connects the basin to the open ocean, is a small and shallow channel (sill depth of 160 m and the narrowest point is 25 km wide). The study area extending from 12.5° N to 30° N and from 32° E to 44° E has an average width of ~220 km and a length of ~2000 km, with a maximum centered depth of 2200 m (Figure 1).

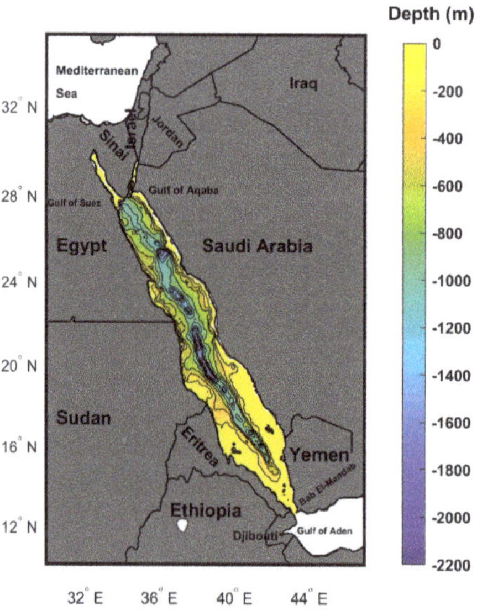

Figure 1. The Red Sea with the main geographic features (e.g., countries). The bathymetry is according to the GEBCO bathymetric dataset (www.gebco.net) (accessed on 15 April 2021), with a minimum depth of 10 m.

2.1. Sea Surface Temperature Climatological Data

The daily Sea Surface Temperature (SST) data used were freely obtained from the Copernicus Marine Environment Monitoring Service website (CMEMS) (https://marine.copernicus.eu (accessed on 20 April 2021)) from 1 January 1982 to 31 December 2019. The data website is https://resources.marine.copernicus.eu/?option=com_csw&view=details&product_id=SST_GLO_SST_L4_REP_OBSERVATIONS_010_011 (accessed on 23 March 2021). The product name is 'SST_GLO_SST_L4_REP_OBSERVATIONS_010_011'. The CMEMS OSTIA (Operational SST and Ice Analysis) reprocessed analysis product is a satellite and in situ foundation SST analysis created by the OSTIA [28]. The SST dataset includes daily gap-free maps of sea surface temperature and ice concentration (known as the L4 product) with a horizontal grid resolution of $0.05° \times 0.05°$ [29]. CMEMS OSTIA data for the Red Sea were extracted from the global data, providing a 16534-point regularly gridded dataset spanning 13,879 days from 1 January 1982 to 31 December 2019.

2.2. Trend and Interannual Variability of SST

The Empirical Orthogonal Functions (EOFs) analysis [30] was used to evaluate the dominant spatiotemporal characteristics of SST variability from 1982 to 2019, utilizing the monthly de-seasonalized and detrended SST dataset to focus on non-seasonal variability. Thus, before decomposing the EOF, the seasonal cycle and linear trends were removed from the monthly means of SST anomalies at each grid cell, and each point time series was divided by its standard deviation to normalize the SST [31,32]. The mean seasonal cycle was estimated by calculating the mean monthly values for each calendar month at each grid point. SST anomalies were computed by removing the historical climatological mean value at each grid point from the SST values at the same location (grid). The strong seasonal signal was removed from the datasets at each grid cell to produce de-seasonalized maps and time series [33]. This was accomplished by subtracting the climatological mean of each calendar month from the corresponding months in all years. The least squares method [34] was used to estimate linear trends in de-seasonalized monthly sea surface temperature anomalies (SSTA). The Modified Mann–Kendall test [35,36] was used to test the statistical significance of the trends. In this study, the significance level was set to $p \leq 0.05$, ensuring that all given trends are statistically significant at least at the 95% confidence level. The error of linear trend at the 95% confidence level was calculated using [34] the formula which was mentioned in [37].

2.3. Marine Heat Waves (MHWs) Calculations

The authors of [5] used the "discrete prolonged anomalously warm-water events in a location" definition of MHWs, using the same daily SST dataset for the 1982–2019 period. The MATLAB toolbox [38] was used to detect periods of time where temperatures were above the 90th percentile threshold compared to the climatology (1982–2019), and with a duration of at least 5 days. The climatological mean has been calculated from the daily Sea Surface Temperature (SST) data from 1 January 1982 to 31 December 2019; the equations of MHW main characteristics are according to [5] and described in [7]. Daily SSTs were spatially averaged over the Red Sea (12.5°–30° N, 32°–44° E) in our study to provide a regional daily SST time series spanning 1982–2019. In order to identify the MHWs in the Red Sea, the MHW definition was then applied to the SST time series.

Every MHW event has a start and end date defined by a mean (i_{mean}) and a maximum intensity (i_{max}) in °C (mean and spatiotemporal maximum temperature compared to the threshold over the MHW event duration (in days)), as described in [39]. The MHW cumulative intensity (i_{cum}) in °C days is defined as the spatiotemporal summation of daily temperature compared to the threshold over the event duration as mentioned by [39]. Following the methodology of [4,40], the annual statistics were computed, including the frequency of events (i.e., the number of discrete events occurring each year). The MHW total number of days in each year can be used to examine how the frequency of occurrences in counts has changed over time, using the metrics for each location [5]. According to [5,7], the main characteristics of MHWs are maximum intensity (i_{max}) in °C, mean intensity (i_{mean}) in °C, and cumulative intensity (i_{cum}) in (°C days):

$$i_{max} = \max \left(T(t) - T_m \left(j \right) \right) \tag{1}$$

$$i_{mean} = \overline{T(t) - T_m(j)} \tag{2}$$

$$i_{cum} = \int_{t_s}^{t_{e-1}} \left(T(t) - T_m \left(j \right) \right) dt \tag{3}$$

where T_m is the climatological OSTIA SST mean calculated over the period from 1 January 1982 to 31 December 2019, to which all values are relative; j is the day of year; and t_s and t_e are the start and end dates of the MHW event, respectively.

3. SST Climatology, Trend, and Interannual Variability

Figure 2a depicts the temporal evolution of the daily mean climatology of Red Sea SST (°C) from 1982 to 2019. The highest temperatures (>29 °C) were in summer and early autumn (i.e., from July to October), while the lowest temperatures (<25 °C) were in winter and early spring (i.e., from January to March). The seasonal temperature difference between summer and winter is about 6 °C, which is consistent with [22]. Figure 2b shows the monthly de-seasonalized SST temporal trend over the study period. The higher de-seasonalized SST values were noticed in the last 4 years of the study period (i.e., 2016–2019). The Red Sea temporal trend was about 0.32 ± 0.03 °C/decade.

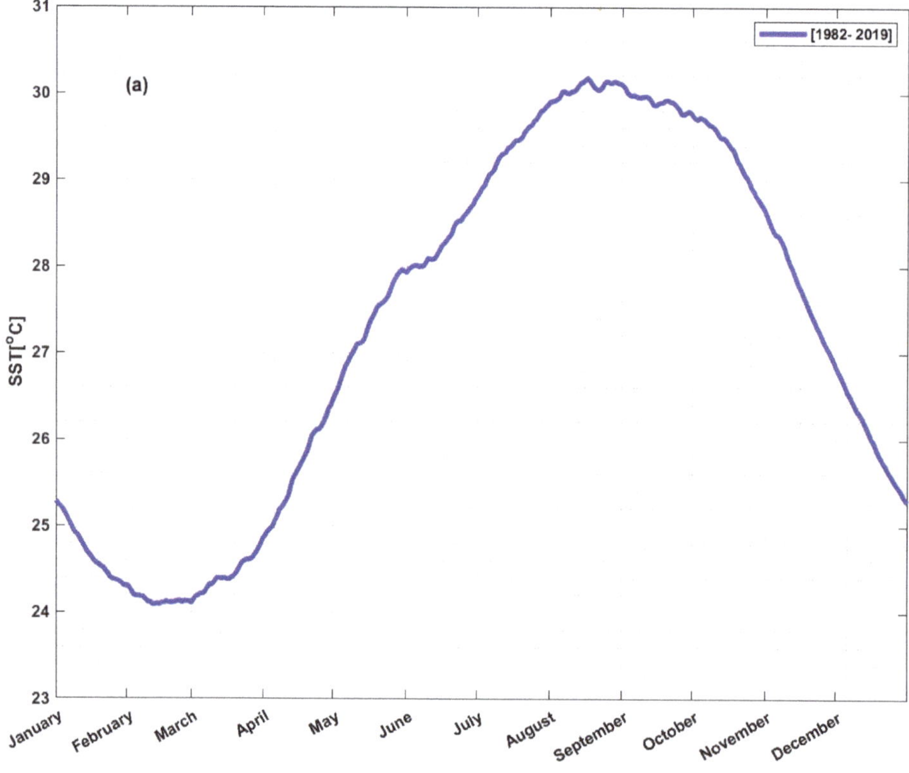

Figure 2. *Cont.*

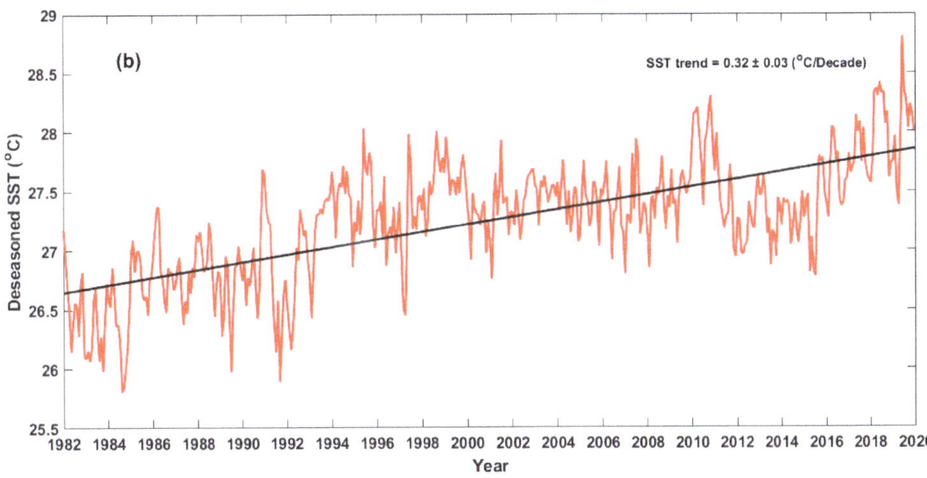

Figure 2. (**a**) The temporal evolution of the daily basin averaged SST (**b**) the de-seasonalized monthly SST time series (denoted by the continuous red line) and the SST trend (denoted by the continuous black line) from 1982 to 2019.

The spatial trend map of the de-seasonalized SSTA from 1982 to 2019 is shown in Figure 3. At the 95% confidence interval, a statistically significant trend was observed over the entire region. The linear trend over the Red Sea is not uniform, varying from 0.1 to 0.47 °C/decade. The basin averaged warming trend rate was about 0.342 ± 0.047 °C/decade. The highest trend was found in the deep area of the Red Sea (i.e., depth >~1200 m), while lower values were found in the Gulfs of Suez and Aqaba and the southern part of the Red Sea.

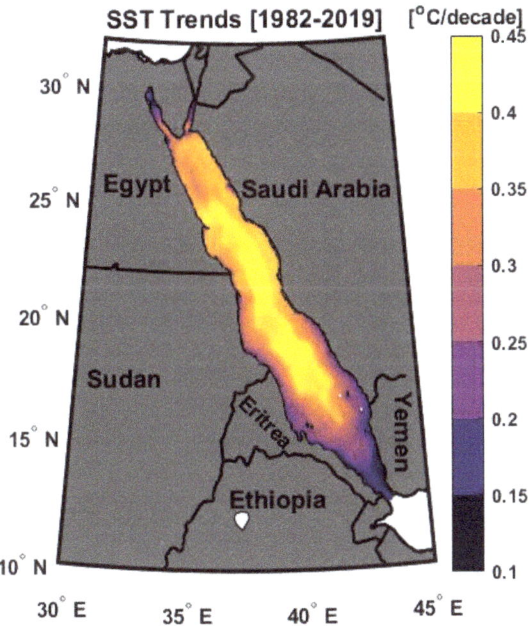

Figure 3. Spatial trend map of the Red Sea SST anomaly from 1982 to 2019, after the seasonal cycle was removed at each grid point.

The first two EOF modes of SST account for 73% of the non-seasonal variance. The spatial pattern of the first EOF (EOF1) (accounting for approximately 52% of the non-seasonal variance) shows a positive anomaly over the entire Red Sea (Figure 4a), showing an in-phase oscillation. The highest variability was found over central and north part of Red Sea, while the lowest variability was detected over the Gulfs of Suez and the Aqaba and the southernmost region of the Red Sea. Strong and continuous northeasterly monsoon winds from the Arabian Sea redirected by mountains into southeasterly winds along the axis of the Red Sea [41] may have a thermal effect that is responsible for the lowest variability over the southern Red Sea. The lower variability in the Gulfs of Suez and Aqaba is due to strong northwesterly winds that penetrate the western edge of the Sinai Peninsula, blow southwards, and sometimes reach the southern end of the Red Sea. Its temporal coefficient PC1 shows a strong interannual variability, reaching its peak in the summers of 1995, 1997, and 2019. The highest negative peaks (i.e., cold periods) were observed in the winters of 1984, 1992, and 2015. The second mode (accounting for approximately 21% of the non-seasonal variance) is out-of-phase variability of the SST in the Red Sea, with positive anomaly values in the south and negative ones in the north. Over the study period of 1982–2019, the temporal coefficient of the second mode (PC2) fluctuates between positive and negative patterns.

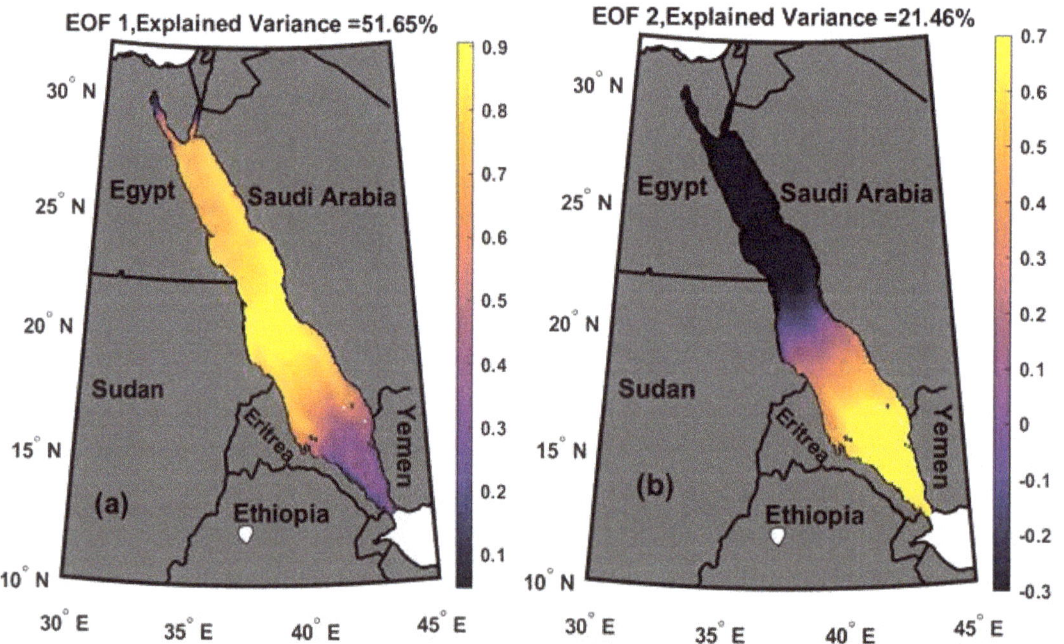

Figure 4. *Cont.*

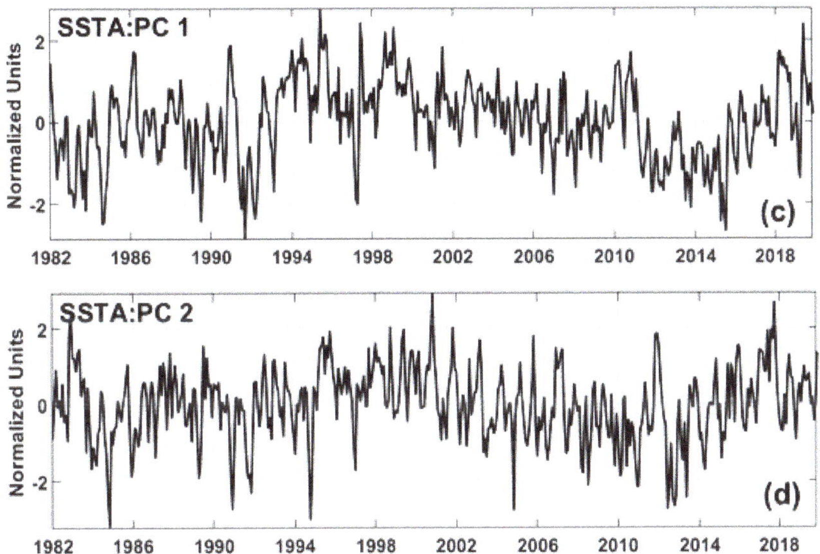

Figure 4. (**a**) Spatial distribution for the first EOF (EOF1); (**b**) Spatial distribution for the second EOF (EOF2); (**c**) Temporal evolution for EOF (PC1) in normalized variance units; (**d**) Temporal evolution for EOF (PC2) in normalized variance units; for the Red Sea SSTA over the period from 1982 to 2019.

4. Main Characteristics and Trends of Marine Heat Waves (MHWs)

The mean and spatial trend of the main characteristics of MHWs (frequency, days, duration, and intensities) over the entire study period are shown in Figure 5. The trend maps are superimposed with non-significant values ($p > 0.05$). Figure 5a demonstrates that the MHW frequency varied from 1 to 2.5 counts, with maximum values of >2 counts found in the coastal areas of Sudan, Eritrea, Yemen and Saudi Arabia. The MHW frequency trend values were statistically significant ($p < 0.05$) over the whole domain. The MHW frequency trend ranged between 0 and 2 counts/decade. The highest MHW frequency trend (>1.8 counts/decade) was detected in the deep area (i.e., depth > 1000 m) between Eritrea and south coast of Saudi Arabia, as shown in Figures 1 and 5b. The MHW total days spatial patterns in the study period (1982–2019) varied between 15 and 30 days, with maximum values of >25 days found inside the Gulfs of Suez and Aqaba (Figure 5c). This could be due to the Gulfs of Suez and Aqaba being well resolved with the high-resolution OSTIA data. In general, the northern Red Sea has higher values of MHW total days than the southern region. The MHW total days trend values were statistically significant ($p < 0.05$) across the whole domain (Figure 5d). We discovered a similarity among the p-values between the MHW frequency and total days trend fields (Figure 5b,d). The MHW mean duration ranged from 5 to 15 days, with maximum values of >12 days north of the Red Sea for the Gulfs of Suez and Aqaba (Figure 5e). The maximum MHW duration spatial trend significant values ($p < 0.05$; >8 days/decade) were found in the middle of the Red Sea (i.e., between ~lat 18° E and 21° E) (Figure 5f). Both the MHW duration and the i_{cum} spatial patterns revealed similarities between the two fields (Figure 5e,g). The MHW i_{cum} values ranged from 10 to 25 (°C days), with maximum values of >22 (°C days) north of the Red Sea, excluding the Gulfs of Suez and Aqaba (i.e., between ~lat 22° E and 28° E) (Figure 5g). The maximum MHW i_{cum} significant trend values ($p < 0.05$; >10 °C days/decade) were discovered in the same areas as the MHW duration trend (Figure 5f,h). The MHW i_{max} showed values of >1.8 °C throughout most of the Red Sea except the southernmost part beside Bab El-Mandab (Figure 5i). The MHW maximum i_{max} significant trend values ($p < 0.05$; >2.5 °C) were found in the middle region of the Red Sea (i.e., between ~lat 19° E and 21° E) and in the north of the Red Sea close to the Gulf of Suez (i.e., between ~lat 26°E and 28° E) off the Egyptian coast (Figure 5j).

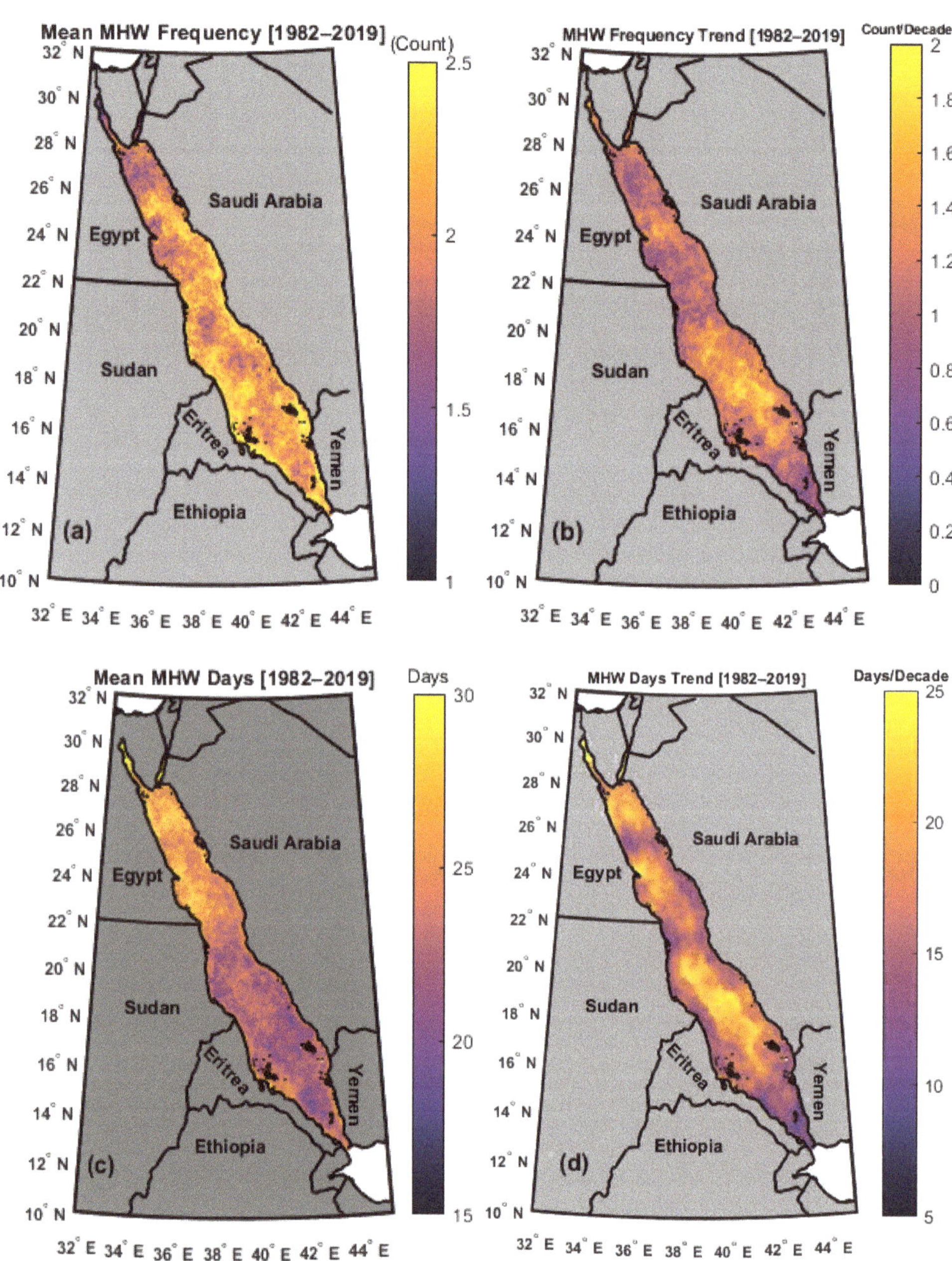

Figure 5. *Cont.*

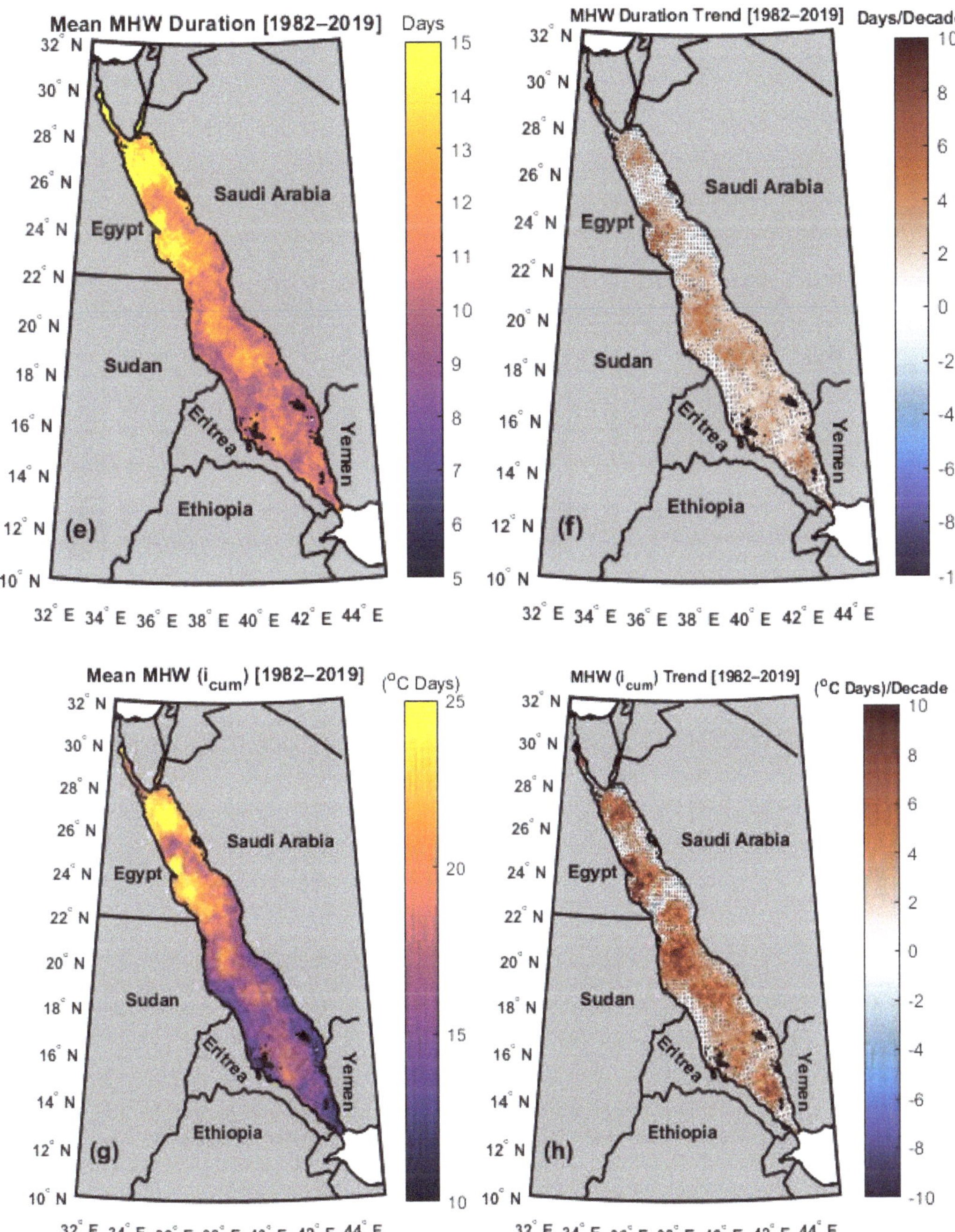

Figure 5. *Cont.*

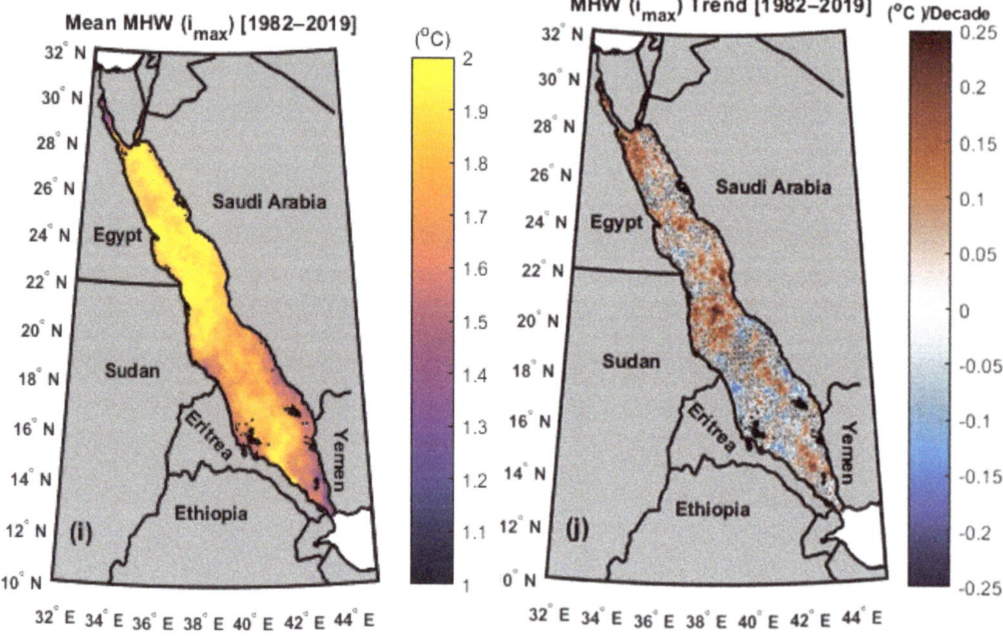

Figure 5. The main characteristics and trends of marine heat waves in the Red Sea over the study period (1982–2019) (**a**) mean MHW frequency (counts); (**b**) MHW frequency trend (counts/decade); (**c**) mean MHW total days (days); (**d**) MHW days trend (days/decade); (**e**) MHW duration (days); (**f**) MHW duration trend (days/decade); (**g**) MHW (i_{cum}) (°C days); (**h**) MHW (i_{cum}) trend (°C days/decade); (**i**) mean MHW (i_{max}) (°C); (**j**) MHW (i_{max}) trend (°C/decade); the black dots indicate that the change is not significantly different ($p > 0.05$).

4.1. Marine Heat Waves (MHWs) Temporal Variation

In this section, the authors present the temporal basin averaged annual MHW main characteristics (frequency, duration, and i_{cum}) from 1982 to 2019. The goal was to investigate temporal changes in Red Sea MHW characteristics such as frequency, duration, and i_{cum}, as well as their relationship to SST (Figure 6). The MHW matrices for the Red Sea region show a clear ascending trend over the last decade, as shown in Figure 6.

From 1982 to 1994, the annual mean MHW frequency was <1, while from 1994 to 2019, the frequency of MHWs increased significantly, reaching a peak in 2018 (7 counts) (Figure 6a). These findings are consistent with the monthly SST time series shown in Figure 2b, from 1982 to 2019. We noticed that no MHWs were detected in 1984. We discovered a strong correlation (R = 0.89) between the annual mean SST and MHW frequency, (Figure 6b).

The results show a general rise in annual MHW duration from approximately 6 days to 21 days in 2018 (Figure 6c), while i_{cum} ranged from 8 °C days to 30 °C days (Figure 6e). The highest annual MHW duration and i_{cum} were found during 2018 (~21 days and 30 °C days). We discovered that the average frequency and duration of MHWs increased by 35% and 67%, respectively, over the last two decades (2000–2019).

The correlation coefficient between MHW duration and annual mean SST was 0.64, while the correlation coefficient between MHW i_{cum} and SST was 0.72.

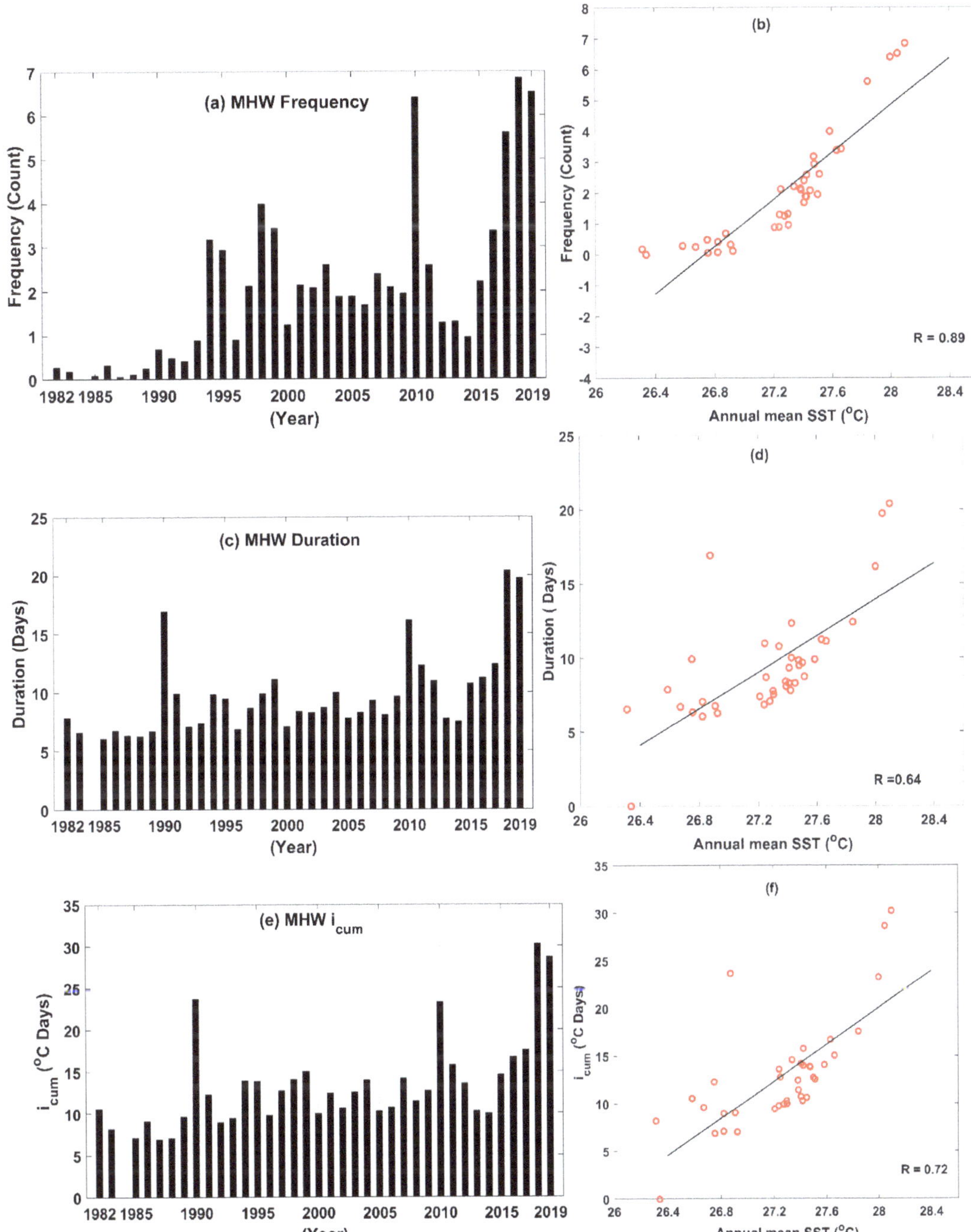

Figure 6. Bars of regional averaged annual means of the MHW (**a**) frequency, (**c**) duration, and (**e**) cumulative intensity (i_{cum}), and scatter plots of annual mean SST and MHW (**b**) frequency, (**d**) duration, and (**f**) cumulative intensity (i_{cum}), over the Red Sea over the period from 1982 to 2019, with the black lines representing the best-fit linear curve.

4.2. Examples of MHW Events

In this section, we show two example MHW events discovered in our MHW analysis between 1982 and 2019. In terms of duration and i_{max}, we chose two MHW events in 2019.

Table 1 shows examples of the Red Sea MHW event characteristics (mean duration, i_{max}, i_{mean}, and i_{cum}) over the study period (1982–2019) with the corresponding MHW_{onset} (start date), MHW_{end} (end date), and longitude and latitude.

Table 1. Examples of the Red Sea MHW event characteristics (mean duration (days), i_{max} (°C), i_{mean} (°C), and i_{cum} (°C days)) over the study period (1982–2019) with the corresponding MHW_{onset}, MHW_{end}, and longitude and latitude (in degrees).

MHW_{onset}	MHW_{end}	Long °E	Lat °N	Mean Duration (Days)	i_{max} (°C)	i_{mean} (°C)	i_{cum} (°C Days)
10 May 2019	31 December 2019	32.375	29.625	236	2.94	1.42	333.94
17 May 2019	21 July 2019	37.875	20.175	66	4.21	2.15	142.06

The long duration event (236 days) with i_{cum} 333.94 °C days was detected in the Gulf of Suez during the period from 10 May 2019 to 31 December 2019, with an i_{max} of 2.94 °C and i_{mean} of 1.42 °C (Figure 7a,b). A high MHW i_{max} event of 4.21 °C was noticed in 2019 off the Sudanese coast (i.e., between ~lat 18.5° E and 20.2° E) (Figure 7c,d) and lasted for 66 days (17 May–21 July) with an i_{mean} of 2.15 °C and i_{cum} of 142.06 °C days.

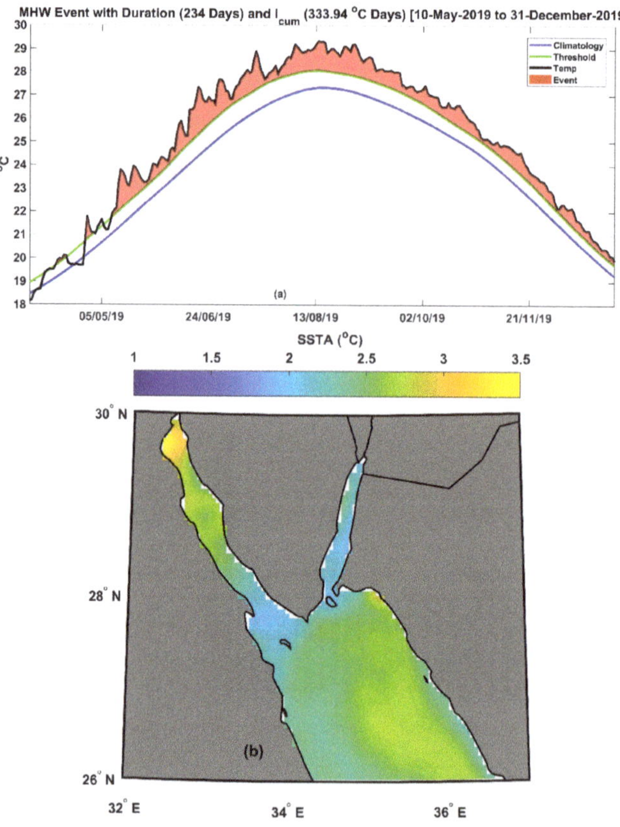

Figure 7. *Cont.*

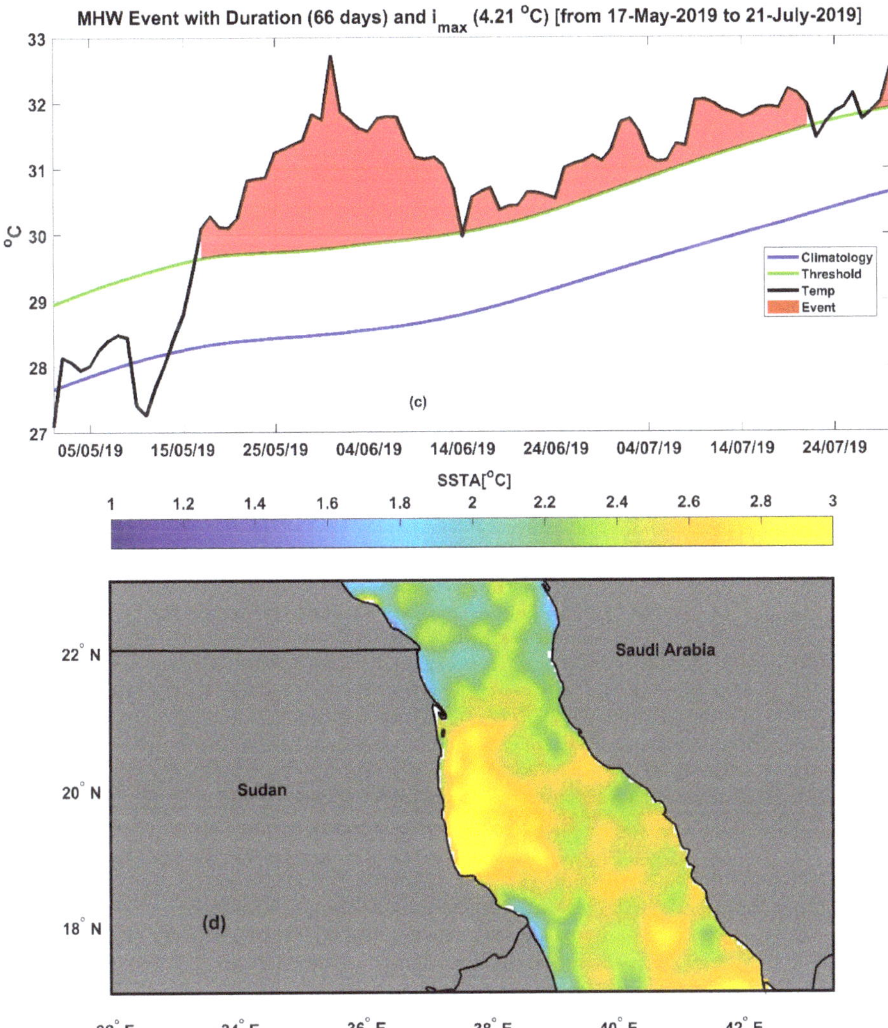

Figure 7. Examples of the Red Sea MHW events over the study period (1982–2019) chosen in the year 2019, The SST climatology from OSTIA SST for the detection of MHWs (blue color), 90th percentile MHW threshold (green color), and SST time series (black color) for each MHW. The area filled with pink color shows the period accompanied with the identified MHW. (**a**) MHW duration event (236 days) (10 May 2019 to 31 December 2019); (**b**) Spatial mean SST anomaly distribution °C (10 May 2019 to 31 December 2019); (**c**) MHW maximum intensity event (i$_{max}$ 4.21 °C) (17 May 2019–21 July 2019); (**d**) Spatial mean SST anomaly distribution °C (17 May 2019–21 July 2019).

5. Discussion

Regarding to our SST climatology and trends analysis (Figures 2 and 3), the highest temperatures (>29 °C, Figure 2a) were found in summer and early autumn. The lowest temperatures (<25 °C, Figure 2a) were recorded in winter and early spring. We found the seasonal temperature difference between summer and winter to be approximately 6 °C (Figure 2a) in agreement with [22]. According to Figure 2b, the Red Sea temporal SST trend was approximately 0.32 ± 0.03 °C/decade from 1982 to 2019, which was higher than [26], possibly due to the use of longer time series and high-resolution OSTIA data.

Many efforts have been made to investigate the Red Sea SST trends using NOAA OI SST coarse resolution data. The authors of [24] showed that August and February are the warmest and coolest months in the Red Sea, with a significant trend of 0.5 °C and 0.3 °C/decade, respectively. According to [15], the overall warming trend of the Red Sea maximum SST was approximately 0.17 °C/decade, while the northern Red Sea alone warmed by between 0.4 and 0.45 °C/decade from 1982 to 2015. The authors of [42] examined the warming of the Red Sea SST from 1982 to 2016 and found an intense warming trend of 0.29 °C/decade. Recently, [26] demonstrated that the SST warming trend ranged from 0.1 °C/decade in the south to 0.4 °C/decade in the north Red Sea between 1982 and 2017. Our analysis revealed a high spatial variability of SST trends over the Red Sea (Figure 3) in the period from 1982 to 2019, varying from 0.10 to 0.47 °C/decade. The basin averaged warming rate was approximately 0.342 ± 0.047 °C/decade. The highest trend was found in the deep area of the Red Sea (i.e., depth >~1200 m), while lower values were found in the Gulf of Suez, the Gulf of Aqaba, and the southern part of the Red Sea.

The EOF analysis (Figure 4) indicated that the highest SST variability was found over the central and northern part of the Red Sea, while the lowest variability was found over the Gulfs of Suez and Aqaba, as well as the southernmost part of the Red Sea. This could be attributed to the persistence of monsoon winds from the Arabian Sea in the southern Red Sea and northwesterly winds from the Sinai Peninsula in the northern Red Sea [41].

In Figure 5, the authors investigated the mean and spatial trend of the main characteristics of MHWs (frequency, days, duration, and intensities) for the Red Sea over the study period (1982–2019). The maximum MHW frequency values of >2 counts were found in the coastal areas of Sudan, Eritrea, Yemen, and Saudi Arabia. The MHW frequency trend values were statistically significant ($p < 0.05$) over the whole Red Sea. According to Figures 1 and 5b, the deep area (i.e., depth > 1000 m) between Eritrea and the south coast of Saudi Arabia has the highest MHW frequency trend (>1.8 counts/decade). The MHW total days' spatial patterns (Figure 5c) show maximum values of more than 25 days in the Gulfs of Suez and Aqaba. This may be due to the fact that the Gulfs of Suez and Aqaba are well resolved using high-resolution OSTIA data. Figure 5b,d show a similarity among the p-values between the MHW frequency and total days trend fields. We found the maximum MHW duration spatial trend significant values ($p < 0.05$; >8 days/decade) to be in the middle of the Red Sea (i.e., between ~lat 18° E and 21° E), as demonstrated by Figure 5f. Maximum MHW i_{cum} values of >22 (°C days) were found in the northern Red Sea, excluding the Gulfs of Suez and Aqaba (i.e., between lat 22° E and 28° E) (Figure 5g). The MHW i_{max} shows high values (>1.8 °C) throughout most of the Red Sea (Figure 5i), with the exception of the southernmost part near Bab El-Mandab. The MHW matrices for the Red Sea region show a clear ascending trend over the last decade (Figure 6). The highest annual MHW duration and i_{cum} were found during the year 2018 (~21 days and 30 °C days). The correlation coefficients between the annual mean SST, MHW duration, and MHW i_{cum} were 0.64 and 0.72, respectively.

The MHW duration event of 236 days was detected in the Gulf of Suez in 2019 from 10 May 2019 to 31 December 2019 (Figure 7a,b). The high MHW i_{max} event of 4.21 °C was discovered in May 2019 off the Sudanese coast (i.e., between ~lat 18.5° E and 20.2° E), (Figure 7c,d).

6. Conclusions

This study provides a comprehensive analysis of SST variability, trends, and spatiotemporal patterns of marine heat waves in the Red Sea from 1982 to 2019. To the authors' knowledge, this is the first study to examine marine heat waves and SST trends in the Red Sea using high-resolution data. A statistically significant SST trend was observed over the Red Sea. The EOF analysis showed that the highest interannual variability was detected over the central and northern part of the Red Sea, while the lowest variability was found over the Gulfs of Suez and Aqaba and the southernmost part of the Red Sea.

We found that all MHW characteristic (frequency, duration, and cumulative intensity (i_{cum})) trends increased over the study period. The highest annual MHW frequencies were detected in the years 2018, 2019, 2010, and 2017. There were no MHWs found during the year 1984. A strong correlation was found between the annual MHW frequency and the annual mean SST. Over the period from 2000 to 2019, we discovered that the average MHW frequency and duration increased by 35% and 67%, respectively. Further research is needed to investigate the causes of MHWs and their effects on marine life in the Red Sea. We believe that our study is just the beginning of a series of studies on this hot topic.

Author Contributions: Conceptualization, O.I., B.M. and H.N.; methodology, H.N., B.M. and O.I.; formal analysis, O.I., H.N. and B.M.; investigation, O.I., B.M. and H.N.; resources, B.M., O.I. and H.N.; data curation, B.M., H.N. and O.I.; writing—original draft preparation, O.I., B.M. and H.N.; writing—review and editing, H.N., O.I. and B.M.; visualization, B.M., O.I. and H.N.; supervision, H.N., B.M. and O.I. All authors have read and agreed to the published version of the manuscript.

Funding: This research received no external funding.

Institutional Review Board Statement: Not applicable.

Informed Consent Statement: Not applicable.

Data Availability Statement: Not applicable.

Acknowledgments: The authors would like to express their appreciation to the two anonymous reviewers for their constructive comments.

Conflicts of Interest: The authors declare no conflict of interest.

References

1. Selig, E.R.; Casey, K.S.; Bruno, J.F. New insights into global patterns of ocean temperature anomalies: Implications for coral reef health and management. *Glob. Ecol. Biogeogr.* **2010**, *19*, 397–411. [CrossRef]
2. Frölicher, T.L.; Laufkötter, C. Emerging risks from marine heat waves. *Nat. Commun.* **2018**, *9*, 1–4. [CrossRef]
3. Smale, D.A.; Wernberg, T.; Oliver, E.C.J.; Thomsen, M.; Harvey, B.P.; Straub, S.C.; Burrows, M.T.; Alexander, L.V.; Benthuysen, J.A.; Donat, M.G.; et al. Marine heatwaves threaten global biodiversity and the provision of ecosystem services. *Nat. Clim. Chang.* **2019**, *9*, 306–312. [CrossRef]
4. Li, Y.; Ren, G.; Wang, Q.; You, Q. More extreme marine heatwaves in the China Seas during the global warming hiatus. *Environ. Res. Lett.* **2019**, *14*, 104010. [CrossRef]
5. Hobday, A.J.; Alexander, L.V.; Perkins, S.E.; Smale, D.A.; Straub, S.C.; Oliver, E.C.J.; Benthuysen, J.A.; Burrows, M.T.; Donat, M.G.; Feng, M.; et al. A hierarchical approach to defining marine heatwaves. *Prog. Oceanogr.* **2016**, *141*, 227–238. [CrossRef]
6. Hobday, A.J.; Oliver, E.C.J.; Gupta, A.S.; Benthuysen, J.A.; Burrows, M.T.; Donat, M.G.; Holbrook, N.J.; Moore, P.J.; Thomsen, M.S.; Wernberg, T.; et al. Categorizing and naming marine heatwaves. *Oceanography* **2018**, *31*, 162–173. [CrossRef]
7. Ibrahim, O.; Mohamed, B.; Nagy, H. Spatial Variability and Trends of Marine Heat Waves in the Eastern Mediterranean Sea over 39 Years. *J. Mar. Sci. Eng.* **2021**, *9*, 643. [CrossRef]
8. Barale, V. The African Marginal and Enclosed Seas: An Overview. *Remote Sens. Afr. Seas Springer Neth.* **2014**, 3–29. [CrossRef]
9. Hoteit, I.; Abualnaja, Y.; Afzal, S.; Ait-El-Fquih, B.; Akylas, T.; Antony, C.; Dawson, C.; Asfahani, K.; Brewin, R.J.; Cavaleri, L.; et al. Towards an end-to-end analysis and prediction system for weather, climate, and Marine applications in the Red Sea. *Bull. Am. Meteorol. Soc.* **2021**, *102*, E99–E122. [CrossRef]
10. Al-Horani, F.A.; Al-Rousan, S.A.; Al-Zibdeh, M.; Khalaf, M.A. The status of coral reefs on the jordanian coast of the gulf of aqaba, red sea. *Zool. Middle East* **2006**, *38*, 99–110. [CrossRef]
11. Eladawy, A.; Nadaoka, K.; Negm, A.; Abdel-Fattah, S.; Hanafy, M.; Shaltout, M. Characterization of the northern Red Sea's oceanic features with remote sensing data and outputs from a global circulation model. *Oceanologia* **2017**, *59*, 213–237. [CrossRef]
12. Sofianos, S.S.; Johns, W.E.; Murray, S.P. Heat and freshwater budgets in the Red Sea from direct observations at Bab el Mandeb. *Deep. Res. Part II Top. Stud. Oceanogr.* **2002**, *49*, 1323–1340. [CrossRef]
13. Johns, W.E.; Sofianos, S.S. Atmospherically forced exchange through the Bab El Mandeb strait. *J. Phys. Oceanogr.* **2012**, *42*, 1143–1157. [CrossRef]
14. Zarokanellos, N.D.; Papadopoulos, V.P.; Sofianos, S.S.; Jones, B.H. Physical and biological characteristics of the winter-summer transition in the Central Red Sea. *J. Geophys. Res. Ocean.* **2017**, *122*, 6355–6370. [CrossRef]
15. Chaidez, V.; Dreano, D.; Agusti, S.; Duarte, C.M.; Hoteit, I. Decadal trends in Red Sea maximum surface temperature. *Sci. Rep.* **2017**, *7*, 8144. [CrossRef]
16. Patzert, W.C. Wind-induced reversal in Red Sea circulation. *Deep Sea Res. Oceanogr. Abstr.* **1974**, *21*, 109–121. [CrossRef]
17. Sofianos, S.S.; Johns, W.E. Observations of the summer Red Sea circulation. *J. Geophys. Res. Ocean.* **2007**, *112*. [CrossRef]

18. Sofianos, S.; Papadopoulos, V.P.; Abualnaja, Y. Investigating the Interannual Variability of the Circulation and Water Mass Formation in the Red Sea. 2014. Available online: https://ui.adsabs.harvard.edu/abs/2014AGUFMOS51A0952S/abstract (accessed on 3 August 2021).

19. Cantin, N.E.; Cohen, A.L.; Karnauskas, K.B.; Tarrant, A.M.; McCorkle, D.C. Ocean warming slows coral growth in the central Red Sea. *Science* **2010**, *329*, 322–325. [CrossRef]

20. Karnauskas, K.B.; Jones, B.H. The Interannual Variability of Sea Surface Temperature in the Red Sea From 35 Years of Satellite and In Situ Observations. *J. Geophys. Res. Ocean.* **2018**, *123*, 5824–5841. [CrossRef]

21. Raitsos, D.E.; Hoteit, I.; Prihartato, P.K.; Chronis, T.; Triantafyllou, G.; Abualnaja, Y. Abrupt warming of the Red Sea. *Geophys. Res. Lett.* **2011**, *38*, 14601. [CrossRef]

22. Berman, T.; Paldor, N.; Brenner, S. Annual SST cycle in the Eastern Mediterranean, Red Sea and Gulf of Elat. *Geophys. Res. Lett.* **2003**, *30*, 1261. [CrossRef]

23. Osman, E.O.; Smith, D.J.; Ziegler, M.; Kürten, B.; Conrad, C.; El-Haddad, K.M.; Voolstra, C.R.; Suggett, D.J. Thermal refugia against coral bleaching throughout the northern Red Sea. *Glob. Chang. Biol.* **2018**, *24*, e474–e484. [CrossRef]

24. Hoegh-Guldberg, O.; Cai, R.; Poloczanska, E.S.; Brewer, P.G.; Sundby, S.; Hilmi, K.; Fabry, V.J.; Jung, S.; Perry, I.; Richardson, A.J.; et al. *SM30-1 30 The Ocean Supplementary Material Coordinating Lead Authors: Lead Authors: Review Editors: To the Fifth Assessment Report of the Intergovernmental Panel on Climate Change*; IPCC: Geneva, Switzerland, 2014; ISBN 9789291691432.

25. Alawad, K.A.; Al-Subhi, A.M.; Alsaafani, M.A.; Alraddadi, T.M. Atmospheric forcing of the high and low extremes in the sea surface temperature over the Red Sea and associated chlorophyll-a concentration. *Remote Sens.* **2020**, *12*, 2227. [CrossRef]

26. Shaltout, M. Recent sea surface temperature trends and future scenarios for the Red Sea. *Oceanologia* **2019**, *61*, 484–504. [CrossRef]

27. Papadopoulos, V.P.; Zhan, P.; Sofianos, S.S.; Raitsos, D.E.; Qurban, M.; Abualnaja, Y.; Bower, A.; Kontoyiannis, H.; Pavlidou, A.; Asharaf, T.T.M.; et al. Factors governing the deep ventilation of the Red Sea. *J. Geophys. Res. Ocean.* **2015**, *120*, 7493–7505. [CrossRef]

28. Good, S.; Fiedler, E.; Mao, C.; Martin, M.J.; Maycock, A.; Reid, R.; Roberts-Jones, J.; Searle, T.; Waters, J.; While, J.; et al. The current configuration of the OSTIA system for operational production of foundation sea surface temperature and ice concentration analyses. *Remote Sens.* **2020**, *12*, 720. [CrossRef]

29. Fiedler, S.; Crueger, T.; D'Agostino, R.; Peters, K.; Becker, T.; Leutwyler, D.; Paccini, L.; Burdanowitz, J.; Buehler, S.A.; Cortes, A.U.; et al. Simulated tropical precipitation assessed across three major phases of the coupled model intercomparison project (CMIP). *Mon. Weather Rev.* **2020**, *148*, 3653–3680. [CrossRef]

30. Thomson, R.E.; Emery, W.J. *Data Analysis Methods in Physical Oceanography*, 3rd ed.; Elsevier Inc.: Amsterdam, The Netherlands, 2014; ISBN 9780123877833.

31. Fenoglio-Marc, L.; Schöne, T.; Illigner, J.; Becker, M.; Manurung, P. Khafid Sea Level Change and Vertical Motion from Satellite Altimetry, Tide Gauges and GPS in the Indonesian Region. *Mar. Geod.* **2012**, *35*, 137–150. [CrossRef]

32. Mohamed, B.; Mohamed, A.; Alam El-Din, K.; Nagy, H.; Elsherbiny, A. Sea level changes and vertical land motion from altimetry and tide gauges in the Southern Levantine Basin. *J. Geodyn.* **2019**, *128*, 1–10. [CrossRef]

33. Skliris, N.; Sofianos, S.; Gkanasos, A.; Mantziafou, A.; Vervatis, V.; Axaopoulos, P.; Lascaratos, A. Decadal scale variability of sea surface temperature in the Mediterranean Sea in relation to atmospheric variability. *Ocean Dyn.* **2011**, *62*, 13–30. [CrossRef]

34. Wilks, D.S. *Statistical Methods in the Atmospheric Sciences*; Academic Press: Cambridge, MA, USA, 2011; ISBN 0123850223.

35. Hamed, K.H.; Ramachandra Rao, A. A modified Mann-Kendall trend test for autocorrelated data. *J. Hydrol.* **1998**, *204*, 182–196. [CrossRef]

36. Wang, F.; Shao, W.; Yu, H.; Kan, G.; He, X.; Zhang, D.; Ren, M.; Wang, G. Re-evaluation of the Power of the Mann-Kendall Test for Detecting Monotonic Trends in Hydrometeorological Time Series. *Front. Earth Sci.* **2020**, *8*, 14. [CrossRef]

37. Mohamed, B.; Abdallah, A.M.; Alam El-Din, K.; Nagy, H.; Shaltout, M. Inter-Annual Variability and Trends of Sea Level and Sea Surface Temperature in the Mediterranean Sea over the Last 25 Years. *Pure Appl. Geophys.* **2019**, *176*, 3787–3810. [CrossRef]

38. Zhao, Z.; Marin, M. A MATLAB toolbox to detect and analyze marine heatwaves. *J. Open Source Softw.* **2019**, *4*, 1124. [CrossRef]

39. Darmaraki, S.; Somot, S.; Sevault, F.; Nabat, P.; Cabos Narvaez, W.D.; Cavicchia, L.; Djurdjevic, V.; Li, L.; Sannino, G.; Sein, D.V. Future evolution of Marine Heatwaves in the Mediterranean Sea. *Clim. Dyn.* **2019**, *53*, 1371–1392. [CrossRef]

40. Oliver, E.C.J.; Donat, M.G.; Burrows, M.T.; Moore, P.J.; Smale, D.A.; Alexander, L.V.; Benthuysen, J.A.; Feng, M.; Sen Gupta, A.; Hobday, A.J.; et al. Longer and more frequent marine heatwaves over the past century. *Nat. Commun.* **2018**, *9*, 1–12. [CrossRef] [PubMed]

41. Langodan, S.; Cavaleri, L.; Vishwanadhapalli, Y.; Pomaro, A.; Bertotti, L.; Hoteit, I. The climatology of the Red Sea—Part 1: The wind. *Int. J. Climatol.* **2017**, *37*, 4509–4517. [CrossRef]

42. Alawad, K.A.; Al-Subhi, A.M.; Alsaafani, M.A.; Alraddadi, T.M. Decadal variability and recent summer warming amplification of the sea surface temperature in the Red Sea. *PLoS ONE* **2020**, *15*, e0237436. [CrossRef]

Journal of
Marine Science and Engineering

Article

Spatial Variability and Trends of Marine Heat Waves in the Eastern Mediterranean Sea over 39 Years

Omneya Ibrahim [1], Bayoumy Mohamed [1,2] and Hazem Nagy [1,3,*]

1   Oceanography Department, Faculty of Science, Alexandria University, Alexandria 21500, Egypt;
    omneya.ibrahim@alexu.edu.eg (O.I.); m.bayoumy@alexu.edu.eg or mohamedb@unis.no (B.M.)
2   Department of Arctic Geophysics, University Centre in Svalbard, 9171 Longyearbyen, Norway
3   Marine Institute, Oranmore, Co Galway, Ireland
*   Correspondence: hazem.nagy@marine.ie or hazem.nagi@alexu.edu.eg; Tel.: +353-89-4985494

Abstract: Marine heatwaves (MHWs) can cause devastating impacts on marine life. The frequency of MHWs, gauged with respect to historical temperatures, is expected to rise significantly as the climate continues to warm. The MHWs intensity and count are pronounced with many parts of the oceans and semi enclosed seas, such as Eastern Mediterranean Sea (EMED). This paper investigates the descriptive spatial variability and trends of MHW events and their main characteristics of the EMED from 1982 to 2020 using Sea Surface Temperature (SST) data obtained from the National Oceanic and Atmospheric Administration Optimum Interpolation ([NOAA] OI SST V2.1). Over the last two decades, we find that the mean MHW frequency and duration increased by 40% and 15%, respectively. In the last decade, the shortest significant MHW mean duration is 10 days, found in the southern Aegean Sea, while it exceeds 27 days off the Israeli coast. The results demonstrate that the MHW frequency trend increased by 1.2 events per decade between 1982 and 2020, while the MHW cumulative intensity (i_{cum}) trend increased by 5.4 °C days per decade. During the study period, we discovered that the maximum significant MHW SST event was 6.35 °C above the 90th SST climatology threshold, lasted 7 days, and occurred in the year 2020. It was linked to a decrease in wind stress, an increase in air temperature, and an increase in mean sea level pressure.

Keywords: marine heatwaves; sea surface temperature; Eastern Mediterranean Sea; maximum intensity; wind stress; mean sea level pressure

Citation: Ibrahim, O.; Mohamed, B.; Nagy, H. Spatial Variability and Trends of Marine Heat Waves in the Eastern Mediterranean Sea over 39 Years. *J. Mar. Sci. Eng.* **2021**, *9*, 643. https://doi.org/10.3390/jmse9060643

Academic Editor: Francisco Pastor Guzman

Received: 21 May 2021
Accepted: 9 June 2021
Published: 10 June 2021

Publisher's Note: MDPI stays neutral with regard to jurisdictional claims in published maps and institutional affiliations.

1. Introduction

In many parts of the world, heatwaves are growing increasingly common and prevalent. It creates significant climatic extremes in oceanic and atmospheric systems, which can have catastrophic ecological impacts and huge socioeconomic effects [1–3]. The study of Marine Heat Waves is important for understanding and predicting climate change and its impact on ocean ecosystems. Marine heat wave (MHW), according to [4], is "a prolonged discrete anomalously warm water event that can be described by its duration, intensity, rate of evolution, and spatial coverage". It was suggested that a linear classification scheme be used by [5], where MHWs are defined based on temperature exceedance from local climatology, which will impact the classification. The probability of the occurrence of MHWs is impacted by SST climatological values as described in [2,6–8]. Changes and variability in SST climatological values, as described in [7,8], will have an impact on the definition and probability of occurrence of MHW. These extreme events lead to a widespread mortality of benthic invertebrates [9] and loss of seagrass meadows [10]. Over the last four decades, the international oceanographic community, regulators and policy makers have paid close attention to the EMED, namely because in this region variations in processes of relevance for the global climate change take place over relatively short space and time scale [11–13].

In the summer of 2003, one of the first-detected MHWs occurred in the Mediterranean which lasted over a month due to an increase of air temperature, a reduction of wind stress,

141

and air-sea exchanges [14–16]. That phenomenon led to an anomalous SST warming in the EMED area during the heatwave of 2003, on the order of +5 °C above climatology [17].

Reference [18] investigated the future evolution (1976–2100) of SST and marine heatwaves in the Mediterranean Sea using a local, climatological 99th percentile threshold based on historical-climate daily SST (1976–2005). The dataset 1982–2012 revealed an annual frequency of 0.8 events per year, with MHWs lasting for a maximum of 1.5 months between July and September, while covering a maximum of 90% of the Mediterranean Sea surface. The longest and most severe event of that period corresponded to the MHW of 2003, with the highest mean intensity and maximum event coverage. The authors have concluded that the MHWs will become stronger and more intense in response to increasing greenhouse gas forcing, particular near the end of the century.

Unfortunately, very few studies on the MHW have been published in the EMED, and the topic is still being debated, as mentioned by [19].

In this study 39 years (1982–2020) of SST data have been analyzed to demonstrate the SST trend and interannual variability based on Empirical Orthogonal Function (EOF). This paper describes the main characteristics of MHWs in the EMED. The authors detect and analyze the most intense MHW events observed in the EMED during the study period, as well as their relationship to atmospheric conditions.

2. Data and Methods of Analysis

The Eastern Mediterranean (EMED) region includes the Ionian and the Levantine basins with two other marginal seas; the Adriatic and the Aegean as shown in Figure 1a. The EMED reaches a depth of over 3000 m off Mersa Matruh at the western part of Egypt and 4000 m to the southeast of Rhodes [20,21] see Figure 1b. The mean depth of the EMED basin is about 2000 m. The shelf is usually extending for no more than 8 km, except off the Nile Delta, where the 200 m contour is traced at 60 km offshore, in the Sicily Strait and the Adriatic Sea. The most important islands in the EMED are Sicily, Malta, Crete, Rhodes, and Cyprus. The study area extending from 30° to 42° N and from 13° to 36.5° E is shown in Figure 1a,b. This area is covering most of the EMED. In this section the authors will describe the data and methods of analysis used to obtain the results.

2.1. Sea Surface Temperature Data

The daily SST data used were obtained from the National Oceanic and Atmospheric Administration Optimum Interpolation Sea Surface Temperature from 1 January 1982 to 31 December 2020 ([NOAA] OI SST V2.1; [22]). The data are freely available on the NOAA website https://www.ncei.noaa.gov/data/sea-surface-temperature-optimum-interpolation/v2.1/access/avhrr/ (accessed on 10 January 2021). This data set is an interpolation of remotely sensed SSTs from the Advanced Very High Resolution Radiometer (AVHRR) imager into a regular grid of 0.25° and daily temporal resolution. AVHRR OI data for the EMED Sea were extracted from the global data, providing a 2624-point regularly gridded dataset spanning 14,245 days from 1 January 1982 to 31 December 2020.

2.2. Trend and Interannual Variabilty of SST

To demonstrate the spatiotemporal variability in SST over the EMED Sea, an EOF is calculated using the singular value decomposition (SVD) approach [23]. EOF analysis extracts the main mode of SST variability, providing us with a spatial pattern and a time-varying index known as the principal component (PC). The EOF analysis was applied to gridded monthly SST anomalies from 1982 to 2020. As described in [24], the SST anomalies were calculated by subtracting the historical climatological mean value at each grid point from the SST values at the same location (grid). Here, we are interested in interannual and longer time scales variability. So, prior to EOF decomposition, the seasonal cycle and linear trends were removed from the monthly means of SST anomalies at each grid point, and then the SST were normalized by dividing each point time series by its standard deviation. The mean seasonal cycle was estimated by calculating the mean monthly values for each calendar month at each grid point from 1982 to 2020. To obtain de-seasoned monthly values, the mean value of each month was subtracted from all corresponding months in all years. Linear trends of de-seasoned monthly SST anomaly are estimated by using the least squares method [25]. To assess the statistical significance of these trends, the original two-tailed modified Mann–Kendall test at 95% confidence interval was used [26,27]. The authors used a thirteen-month running mean to remove the higher frequency variability in order to highlight the interannual variability, as described in [28]. Refs. [29,30] provide examples of similar data filtration methods.

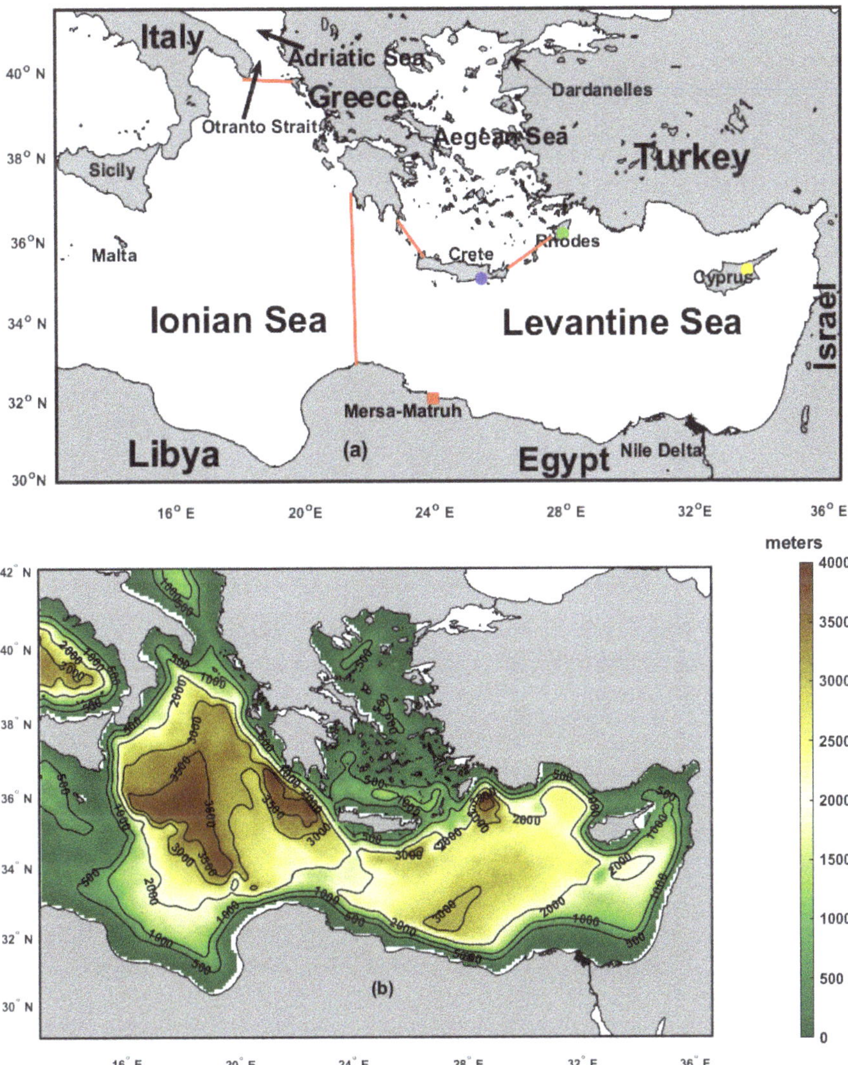

Figure 1. (a) The Eastern Mediterranean Sea with the main geographic features (e.g., Basins) where, the solid red square is the position of Mersa Matruh City, the solid blue circle is the position of Crete, the solid green circle is the position of Rhodes, the solid yellow circle is the position of Cyprus and the solid continuous red lines delineate the boundaries between EMED basins. **(b)** Bathymetry according to GEBCO bathymetric dataset (https://www.gebco.net/ (accessed on 20 March 2021)) with 10 m as the minimum depth.

It is crucial for our study to determine the warmest period from the whole climatological SST time series, NOAA OI SST V2.1. To reach our goal we divided the SST time series into four periods, (P_1, 1 January 1982–31 December 1990), (P_2, 1 January 1991–31 December 2000), (P_3, 1 January 2001–31 December 2010) and (P_4, 1 January 2011–31 December 2020). The chosen of the first period P_1 is based on satellite NOAA OI SST V2.1 data availability. Then daily climatology was calculated from the P_1, P_2, P_3 and P_4 years employed in the analysis and smoothed with a monthly centred-moving mean.

2.3. Marine Heat Waves (MHWs) Calculations

The authors used [5] MHWs, definition, as a discrete prolonged anomalously warm-water events in a location, using the same daily SST database for the 1982–2020 period. The MATLAB toolbox [31] has been used to identify periods of time where temperatures were above the 90th percentile threshold relative to the climatology, and for a duration of at least 5 days. The authors identified the start and end dates and calculated the duration and intensity for each MHW event according to [5]. The climatological mean has been calculated from the daily Sea Surface Temperature (SST) data from 1 January 1982 to 31 December 2020 ([NOAA] OI SST V2.1; [22] following the [5] formula,

$$T_m(j) = \sum_{y=y_s}^{y_e} \sum_{d=j-5}^{j+5} \frac{T(y,d)}{11\,(y_e - y_s + 1)} \tag{1}$$

where, T_m is the climatological SST mean in °C calculated over a period from 1 January 1982 to 31 December 2020, to which all values are relative, j is the day of year, y_s and y_e are the start and end of the climatological base period respectively, and T is the daily SST on day d of year y. Then the 90th percentile threshold based on SST climatology period, T_{90} (j) in °C are calculated according to [5] based on climatological SST mean (X) in °C over a period from 1 January 1982 to 31 December 2020 as follow:

$$T_{90}\,(j) = P_{90}\,(X) \tag{2}$$

According to [5] definition, MHWs occur when SSTs exceed the threshold, defined as the 90th percentile of SST variations (T_{90} (j)) based on a 39-year climatological period (1982–2020), for at least five consecutive days. At each location and for each MHW, we calculated the event duration (time between start and end dates) and mean cumulative intensity. Associated annual statistics were then calculated, including the frequency of events (i.e., the number of discrete events occurring in each year),

$$t_e - t_s \geq 5 \tag{3}$$

where t_e, t_s: dates on which a MHW begins and ends. And D is the MHW Duration in Days that temperature exceeds the threshold, $D = t_e - t_s$.

Where P_{90} is the 90th percentile and

$$X = \{T(y,d)\,|\,y_s <= y <= y_e, j - 5 <= d <= j + 5\} \tag{4}$$

Then, the MHWs main characteristics are (i_{max}) in °C, mean intensity (i_{mean}) in °C and (i_{cum}) in (°C Days):

$$i_{max} = \max\,(T(t) - T_m\,(j)) \tag{5}$$

$$i_{mean} = \overline{T(t) - Tm(j)}\;\textbf{During the MHW event}_s \leq j \leq t_e \text{ and } j(t_s) \leq j \leq j(t_e) \tag{6}$$

$$i_{cum} = \int_{t_s}^{t_e-1} (T(t) - T_m(j))\,dt \tag{7}$$

i_{cum}: sum of daily intensity anomalies.

According to [5] during the MHW event:

$$T(t) > T_{90} (j), t_s \text{ is the time, } t,$$

while during time $(t - 1)$ presents no MHW event

$$T(t - 1) < T_{90} (j), t_e \text{ is the time, } t, \text{ where } t_e > t_s$$

$$T(t) < T_{90} (j)$$

2.4. Atmospheric Variables Data

Atmospheric variables were obtained from the European Centre for Medium-Range Weather Forecasts (ECMWF) ERA5 [32]. Winds at 10 m elevation (U_{10} and V_{10}), air temperature at 2 m elevation, and pressure at mean sea level were used. The dataset has $0.25°$ spatial resolution with hourly temporal step. Daily means were computed by averaging hourly data, and a daily climatology was built in the same manner as for SST, using the same period from 1982 to 2020. The wind stress components in east and north directions (τ_{wx}, τ_{wy}) were calculated according to [33] in (N/m^2),

$$\tau = C_D \, \rho_a \, |W| * W \tag{8}$$

where $\rho_a = 1.2$ (kg/m^3) is the density of air, C_D is the drag coefficient in Equation (8) is computed according to [34] while (W) is the wind speed component (m/s). The authors have computed the wind stress components to track 14–20 May 2020 MHW events which have the maximum intensity over the whole study period (1982–2020).

3. Description of the SST Data (1982–2020)

Figure 2 shows the temporal evolution of daily mean climatology SST ($°C$) for P_1, P_2, P_3 and P_4. The figure clearly shows that P_1 has the lowest climatological SST values, with a maximum of about 26.01 $°C$, a minimum of about 15.07 $°C$, and a mean value of 19.98 $°C$ (see Table 1). During P_2, the highest climatological SST was approximately 26.38 $°C$, and the lowest was approximately 15.25 $°C$, with a mean of 20.12 $°C$. The maximum, minimum, and mean statistical values for P_3 were 26.52 $°C$, 15.44 $°C$, and 20.48 $°C$, respectively. During P_4, the maximum value reached 27.07 $°C$ and the minimum was 16.76 $°C$ with a mean value of 20.92 $°C$.

Table 1. The basin average, minimum and maximum climatological SST $°C$ during P_1, P_2, P_3 and P_4.

Decade	Mean	Minimum	Maximum
P_1 (1982–1990)	19.98	15.07	26.01
P_2 (1991–2000)	20.12	15.25	26.38
P_3 (2001–2010)	20.48	15.44	26.52
P_4 (2011–2020)	20.92	16.76	27.07

In summary, the maximum values over the four decades were detected during P_4 in agreement with [19,35,36]. Figure 2 shows that SST values are gradually increasing from one decade to the next. The EMED basin average climatological SST values in Table 1 show that P_4 was the warmest decade by about 0.94 $°C$ within the entire 1982–2020 study period.

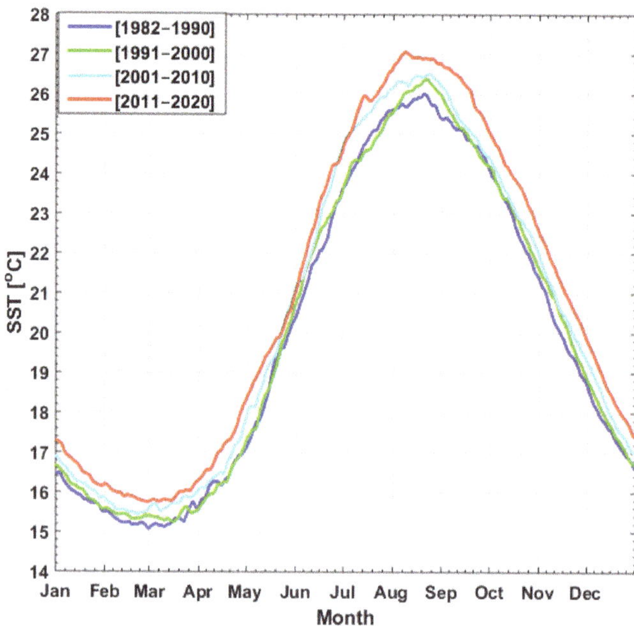

Figure 2. Temporal evolution of the basin averaged plotted SST in (°C) where P_1 (1982–1990) in continuous blue line, P_2 (1991–2000) in continuous green line, P_3 (2001–2010) in continuous cyan line and P_4 (2011–2020) in continuous red line.

The authors used anomaly maps to clearly identify cooling and warming regions over the aforementioned time periods.

As shown in Figure 3a, the P_1 SST anomaly map shows a cooling range of -1.00 to -0.60 °C that covers the entire EMED. The P_2 east of Crete is cooling to cc. -0.25. (see Figure 3b). Figure 3c shows that the cooling rate for P_3 has decreased. The cooling range was between -0.50 to -0.10 °C. The P_4 anomaly patterns revealed a warming of about 0.10 to 0.50 °C, over the northern EMED basin except some parts of the north Ionian Sea. This region, which is known as a Bimodal Oscillating System (BiOS), is responsible for decadal reversals of the Ionian basin-wide circulation as described by [37–41]. It is a decadal circulation periodic reversal region from cyclone to anticyclone and vice versa as described by [41]. The BiOS is regarded as a pattern of EMED climate change as mentioned by [40]. In general, the northern EMED basin's SST anomaly for P_4 is greater (by 0.25 to 1.00 °C) than the southern one. As shown in Figure 3d, the maximum SST anomaly for P_4 was 0.50 °C in the northwest Aegean Sea, while the minimum value was -0.50 °C along the southwestern coast of EMED. This may be attributed to the fact that, the northern part of the EMED basin is occupied by a cyclonic motion such as Rhodes cyclonic gyre as mentioned in [42], while the southern part is dominated by anticyclonic gyres such as Mersa Matruh and Shikmona anticyclonic gyres as described in [37–43]. Hence, warming in anticyclonic areas is less than in the cyclonic ones due to strong stratification in anticyclonic regions [44,45].

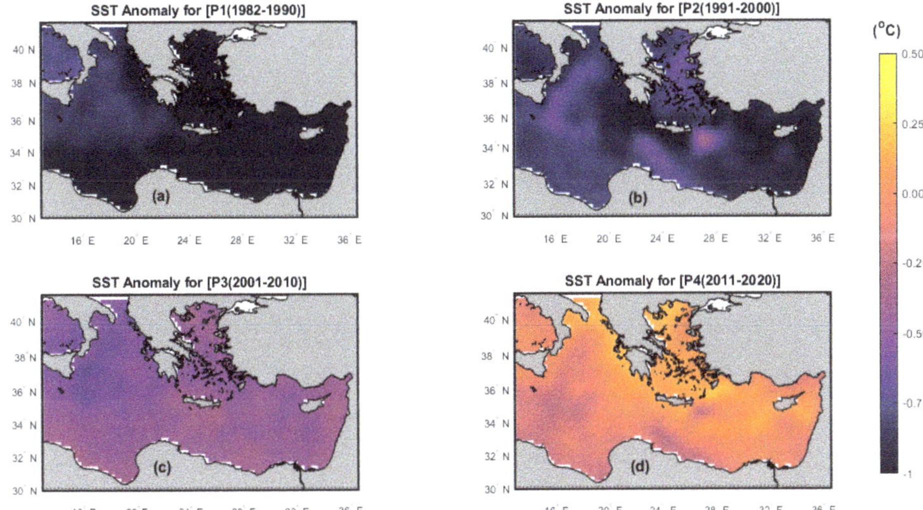

Figure 3. The EMED SST climatological anomaly maps where (**a**) P_1 (1982–1990), (**b**) P_2 (1991–2000), (**c**) P_3 (2001–2010), and (**d**) P_4 (2011–2020). Anomaly maps are calculated by subtracting the climatological mean value over (1982–2020) from each period value (P_1, P_2, P_3 and P_4) at the same location (grid).

Overall, the patterns of SST anomaly for the four periods show that P_4 is the warmest period across the study area. These information highlight some specific hotspot areas with the highest SST anomaly during P_4, such as the Aegean, north Levantine, and north Ionian.

Trend and Inter-Annual Variability of SST

Figure 4 depicts a linear trend map after removing SST seasonality from 1982 to 2020. The original two-tailed modified Mann-Kendall test [26,27] was applied to the estimated trend at each grid point, the results indicate a statistically significant trend at the 95% confidence interval was observed over the whole region. High spatial variability of SST linear trend was observed over the EMED, ranging from 0.1 to 0.5 °C per decade. The basin average warming rate was about 0.33 ± 0.04 °C per decade. The higher values were found in the Aegean and Levantine Basin, while the lower values were observed in the Ionian and Tyrrhenian Sea.

The first three EOF modes of SST anomaly explain about 71% of the overall deseasoned variance of the data. We only discuss the first two modes of EOF since the third mode describes less than 5% of the variance.

The spatial pattern of the first EOF1 mode, which accounts for approximately 50% of the overall deseasoned variance, is a positive anomaly over the entire region of the EMED (Figure 5a), suggesting an in-phase oscillation of the entire basin around the steady-state mean. The Ionian Sea and the Mersa Matruh gyre in the west of the Egyptian coast had the highest SST variability, while the Aegean Sea had the lowest variability, owing to the presence of strong and persistent Etesian winds over the Aegean Sea [35] and the input of Black Sea cold water inflow into the Aegean Sea [28]. The first temporal mode (PC1) reached its peak in the summer of 2003 (Figure 5b), which was the warmest summer over the study period, due to the strongest heat waves that hit the northern Europe. The highest negative peaks were observed in the winters of 1987 and 2019 in agreement with [46]. The lower peak values (i.e., cold periods) in the 1990s could be attributed to Mount Pinatubo's eruption in the Philippines, which caused a cold air-temperature anomaly throughout the Middle East during the winter of 1992 As mentioned by [47], The spatial pattern of the second EOF2 mode (Explains 16% of the total variance) demonstrated a dipole in the study area (Figure 5c), with positive anomalies in the Ionian Sea and negative ones in the Levantine and Aegean basins. The temporal coefficient of the second mode (PC2) alternates between positive and negative patterns over the study period 1982–2020.

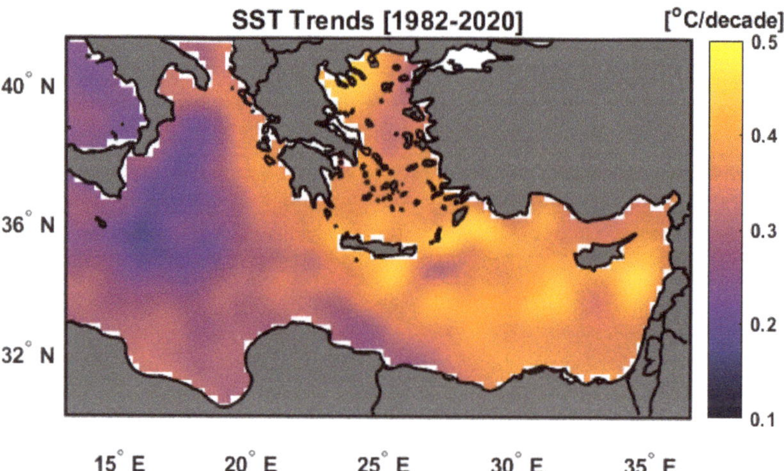

Figure 4. Spatial trend map of SST anomaly over the EMED from AVHRR during 1982–2020, after the seasonal cycle was removed at each grid point.

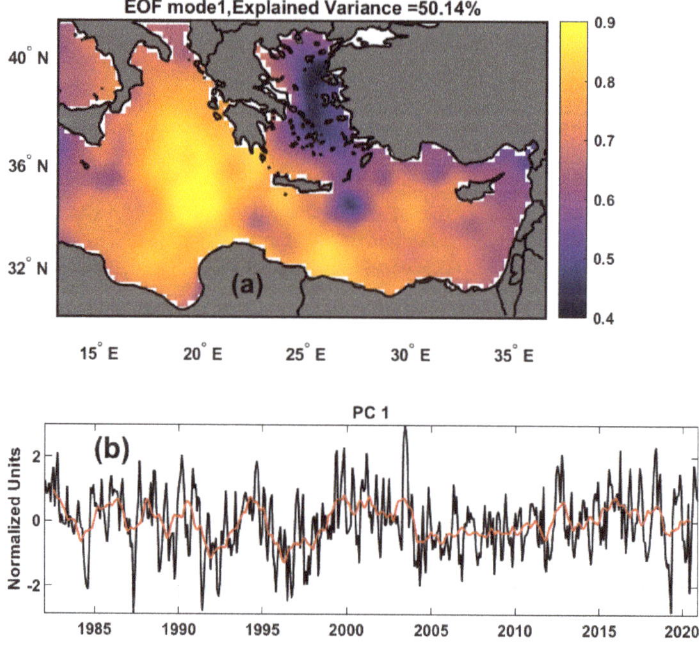

Figure 5. *Cont.*

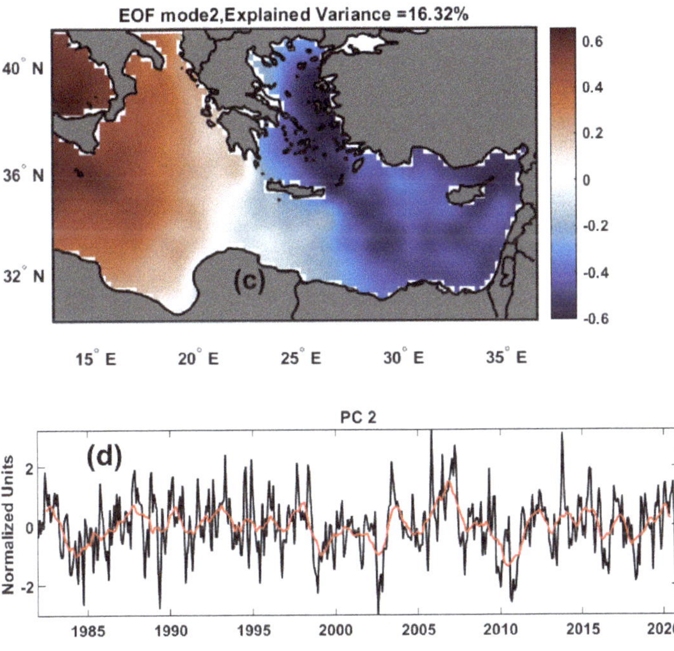

Figure 5. (**a**) Spatial distribution for the principal (EOF) component (PC1). (**b**) Temporal evolution for EOF (PC1) in normalised variance units. (**c**) Spatial distribution for the second (EOF) component (PC2). (**d**) Temporal evolution for EOF (PC2) in normalised variance units; for the EMED SST anomaly covering the period from 1982 to 2020. In (**b**,**d**), the deseasoned monthly PCs time series is shown in continuous black lines, and the low passed using a 13-month running mean is shown in continuous red lines.

4. Spatial Description and Trends of the MHWs Main Characteristics from 1982 to 2020

Figure 6a–h depicts the spatial distribution and trends per decade of the EMED MHWs main characteristics, frequency, duration, (i_{cum}), and (i_{max}), which are overlaid with no significant values ($p > 0.05$) for the entire study period.

There was a clear decadal increase in MHW frequency in all EMED locations as demonstrated in Figure 6a,b. The frequency ranged from 1–2.5 counts with a maximum significant ($p < 0.05$) values > 2 events found in most of the Aegean and Levantine Seas, south of Crete and in front of the eastern coast of Libya. As shown in Figure 6b, the MHW frequency trend values were positively significant ($p < 0.05$) and ranged between 0 and 2.5 counts per decade. There were no significant ($p > 0.05$) changes in MHW frequency southeast of Crete, in front of Mersa Matruh city (see Figure 1a), and north of the Ionian Sea region (BiOS), which was previously described as a decadal reversal region in Section 3. The patterns of i_{cum} (trend) and mean duration (trend) are nearly identical, as shown in Figure 6c–f. The MHW mean duration increased significantly ($p < 0.05$) with values > 17 days with a trend of >10 days per decade in front of the Egyptian coast in the southeast of the Levantine Sea and off the Israeli coast, Figure 6d. No significant values ($p > 0.05$) are found in MHW duration at southeast of Crete, off Mersa Matruh city and the north Ionian Sea region. The maximum significant ($p < 0.05$) MHW i_{cum} values found in the Levantine Sea > 30 °C Days with a trend >15 °C Days per decade (Figure 6e,f). The MHW i_{max} patterns reveal maximum significant values > 1.5° with a trend > 0.25 °C throughout most of the Levantine Sea with an exception of south east corner and off Mersa Matruh city (see Figure 6g,h).

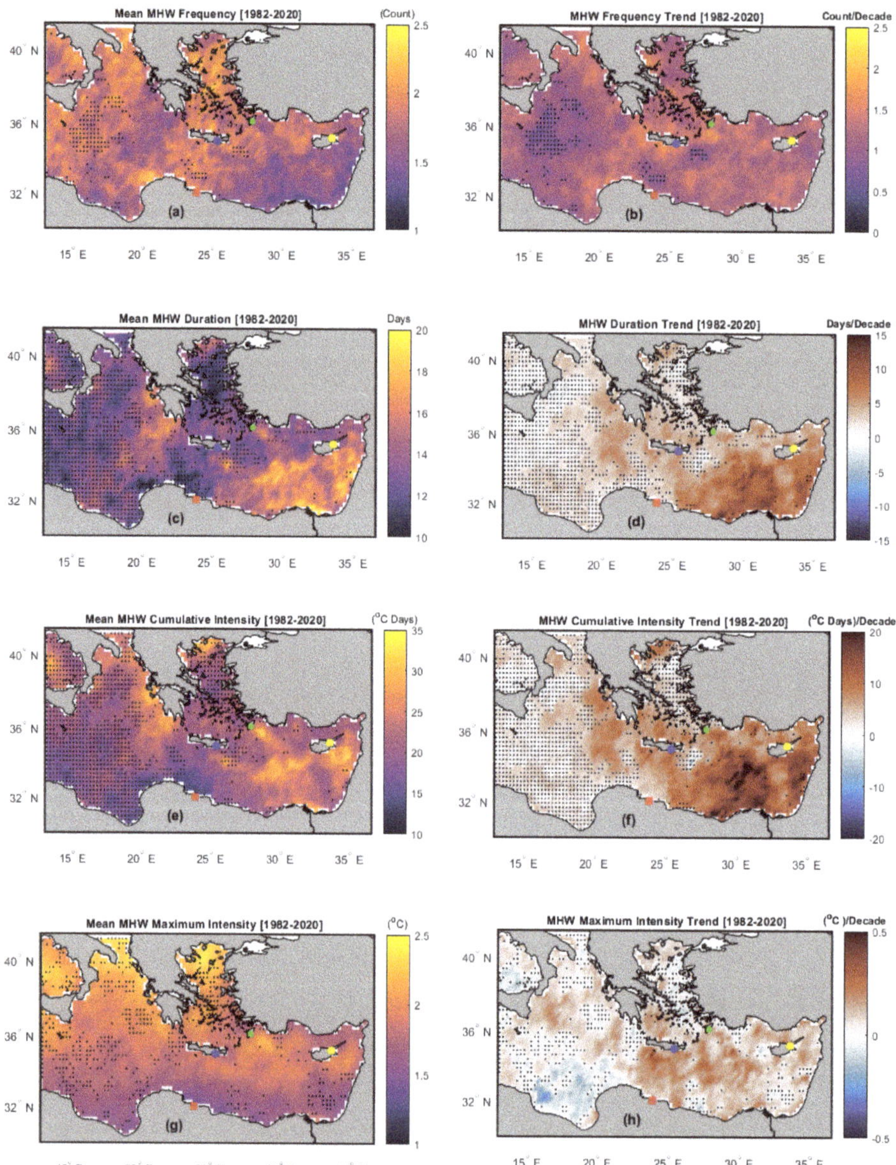

Figure 6. The Marine heatwave properties over the 1982–2020 period, averaged time series of (**a**) mean MHW frequency (Count), (**b**) MHW frequency trend (Count/Decade), (**c**) mean MHW duration (Days), (**d**) MHW duration trend (Days/Decade), (**e**) mean MHW (i_{cum}) (°C Days), (**f**) MHW (i_{cum}) trend (°C Days/Decade), (**g**) mean MHW (i_{max}) (°C), (**h**) MHW (i_{max}) trend (°C/Decade), and the black dots indicate the change is not significantly different ($p > 0.05$). Where, the solid red square is the position of Mersa Matruh City, the solid blue circle is the position of Crete, the solid green circle is the position of Rhodes, and the solid yellow circle is the position of Cyprus.

The authors found a strong relationship between the MHWs mean frequency and mean duration at each location over the study period (see, Figure 6a,c). The spatial pattern of mean MHW frequency and duration was highly negatively correlated ($r = -0.76$), which means the frequency was low where duration was long, and vice versa, example of similar analysis can be found in [3].

4.1. Description of the MHWs Mean Frequency Fields (1982–2020)

Figure 7a–d shows the spatial distribution of the EMED MHWs mean frequency events for the study periods. The P_1 mean MHW events are significantly < 2 counts over the whole EMED domain, Figure 7a. While during P_2 the MHW events have significantly increased up to 3 counts, Figure 7b. We found during P_3 that the Levantine basin has more significant ($p < 0.05$) MHW events (>3) than the Ionian and Aegean Seas, see Figure 7c. P_4 shows that most significant ($p < 0.05$) MHW events > 4 when compared to the other periods, Figure 7d.

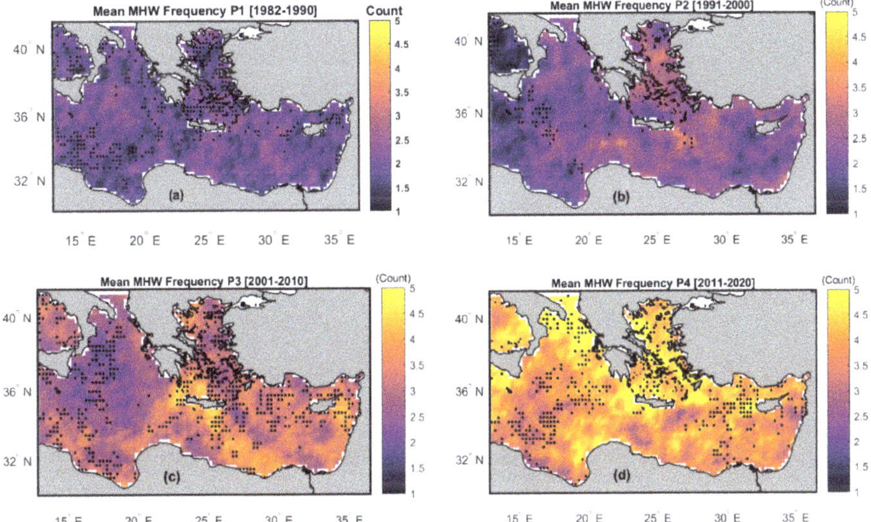

Figure 7. The EMED MHWs mean frequency (Count) for (**a**) P_1 (1982–1990), (**b**) P_2 (1991–2000), (**c**) P_3 (2001–2010) and (**d**) P_4 (2011–2020), and the black dots indicate the change is not significantly different ($p > 0.05$).

4.2. Description of the MHWs Mean Duration (Days) Fields (1982–2020)

Figure 8a–d describes the spatial distribution of the EMED MHWs mean Duration (days) for the different periods, (P_1, P_2, P_3 and P_4). The spatial field for MHW mean duration over P_1 is mostly nonsignificant ($p > 0.05$) and less than 10 days for most of the domain. With an exception in the North Ionian area where the duration is significantly more than 18 days as shown in Figure 8a. For the most of the EMED domain, there was no significant difference in MHW mean duration ($p > 0.05$) during P_2. The MHW mean duration for P_2 ranges from 6.71 days to 20.53 days. We noticed during P_2 a high significant ($p < 0.05$) MHW mean duration value > 20 days located off South of Crete, see Figure 8b. P_3 and P_4 show an increase of significant MHW mean duration values throughout the EMED domain. For P_3, the MHW mean duration ranges from 9.07 days to 21.36 days, Figure 8c. The MHW mean duration pattern over the last decade shows greater significant values in the Levantine Sea. The least significant ($p < 0.05$) MHW mean duration is 10 days found in the South of the Aegean Sea, while it significantly exceeds 27 days off the Israeli coast as shown in Figure 8d. Finally, the findings revealed that the mean duration of MHWs has increased rapidly over the last two decades.

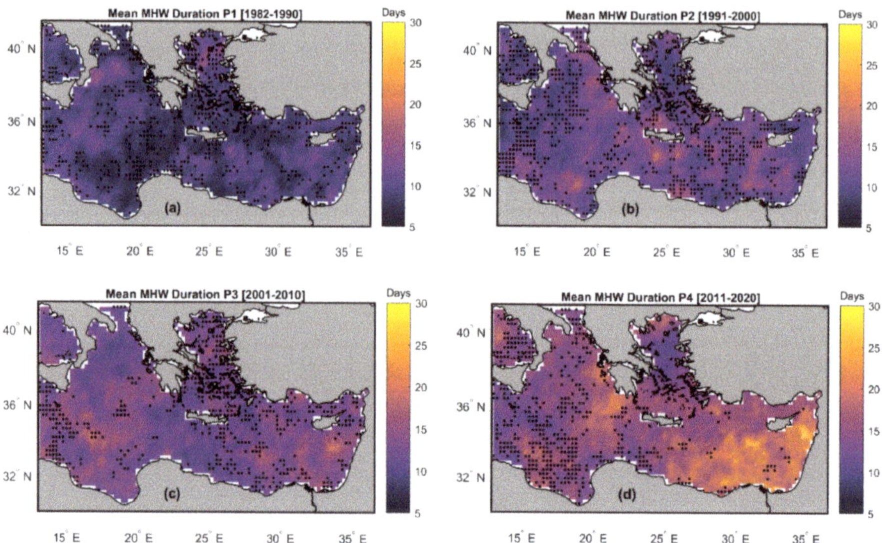

Figure 8. The EMED MHWs mean duration (days) for (**a**) P_1 (1982–1990), (**b**) P_2 (1991–2000), (**c**) P_3 (2001–2010) and (**d**) P_4 (2011–2020), and the black dots indicate the change is not significantly different ($p > 0.05$).

4.3. Description of the MHWs (i_{cum}) Fields (1982–2020)

Figure 9a–d presents the spatial distribution of the EMED MHWs mean cumulative intensity (°C days) for the different periods, (P_1, P_2, P_3 and P_4). The MHW i_{cum}, defined as the integral of intensity over the duration of the event [5], represents the MHW severity as mentioned by [48]. The mean spatial pattern for P_1 in Figure 9a shows non-significant values for most of the EMED basin. The highest P_1 significant ($p < 0.05$) value is >30 (°C days) and detected to the south of Italy in the Ionian Sea. Figure 9b demonstrates that the significant ($p < 0.05$) MHW i_{cum} values have increased for P_2. It ranges from 9.60 to 32.63 (°C days). The maximum MHW significant i_{cum} values for P_3 is >30 (°C days) and are located to the South of Cyprus and in the middle of the Ionian Sea, Figure 9c. The map of P_4 shows an increase of significant severity, which varies between 14.29 (°C days) and 39.51 (°C days) in the EMED domain. The highest significant MHW i_{cum} values > 35 (°C days) can be detected off the Israeli coast near the Shikmona anti-cyclonic gyre location as shown in Figure 9d. The Shikmona anti-cyclonic eddy was located the area between 34° E and 35° E longitudes and 33.5° N and 34.5° N latitudes [43,49,50].

4.4. Description of the MHWs (i_{max}) Fields (1982–2020)

Figure 10a–d shows the spatial distribution of the EMED MHWs i_{max} (°C) for the study periods. During P_1, the maximum significant ($p < 0.05$) mean MHW i_{max} was found throughout most of the Ionian Sea with values > 2 °C, Figure 10a. The significant MHW i_{max} values during P_2 vary between 1.33 °C at southeast Levantine basin and 2.45 °C at northeast Ionian basin near the Otranto strait, Figure 10b. The highest significant value for P_3 is 2.5 °C and is found north of the Aegean Sea, while the lowest significant value (1.31 °C) is found southeast of the Levantine Sea in front of the Egyptian coast. Among the entire study period, the MHW i_{max} spatial pattern for P_4 has the highest significant ($p < 0.05$). The MHW i_{max} values are significant during P_4 throughout the whole Levantine basin, Figure 10d. Figure 10d shows that the significant MHW i_{max} values range from 1.42 °C off Mersa Matruh to more than 2.5 °C in the northeast Aegean basin.

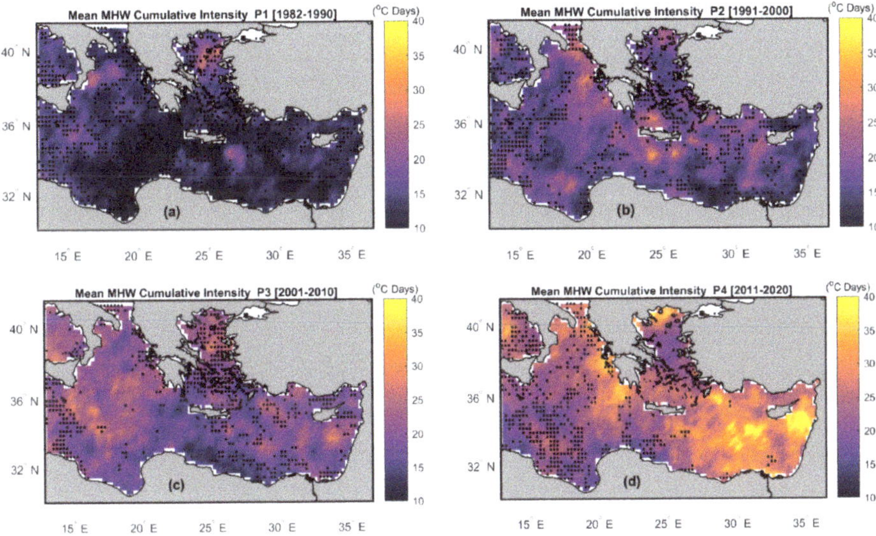

Figure 9. The EMED MHWs mean cumulative intensity (°C days) for (**a**) P_1 (1982–1990), (**b**) P_2 (1991–2000), (**c**) P_3 (2001–2010) and (**d**) P_4 (2011–2020), and the black dots indicate the change is not significantly different ($p > 0.05$).

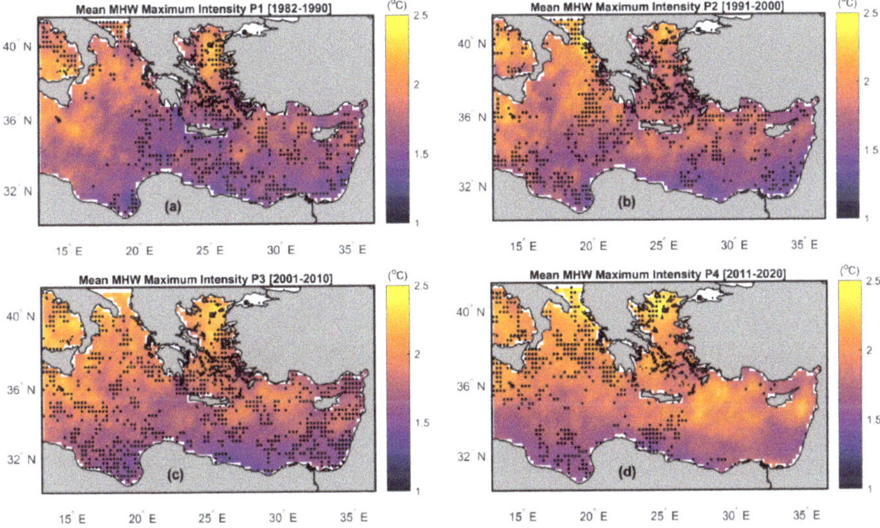

Figure 10. The EMED MHWs i_{max} (°C) for (**a**) P_1 (1982–1990), (**b**) P_2 (1991–2000), (**c**) P_3 (2001–2010) and (**d**) P_4 (2011–2020), and the black dots indicate the change is not significantly different ($p > 0.05$).

5. Extreme MHWs Events in the EMED over 1982–2020

Table 2 shows the significant EMED extreme MHW events characteristics (mean duration, i_{max}, i_{mean}, and i_{cum}) over the study period (1982–2020) with the corresponding MHW_{onset}, MHW_{end}, Longitude and Latitude. The maximum MHW i_{mean} was 3.92 °C in 2013, Figure 11a and lasted for 15 days (24 April–08 May) with an i_{max} of 5.06 °C and i_{cum} of 58.77 °C days. The most severe event (i_{cum} of 319.65 °C days) was still ongoing at the time of the analysis with 126 days, i_{max} of 3.41 °C and i_{mean} of 2.54 °C, Figure 11b. The most intense event occurred in 2020, with an i_{max} temperature of 6.35 °C and a duration of 7 days (this event will be discussed in detail in the following subsection). In general, the most extreme MHW significant events were mostly detected in 2020.

Table 2. The EMED extreme significant ($p < 0.05$) MHW events characteristics (mean duration (Days), i_{max} (°C), i_{mean} (°C), and i_{cum} (°C days)) over the study period (1982–2020) with the corresponding MHW_{onset}, MHW_{end}, Longitude and Latitude in degrees where the highest events are highlighted in yellow color.

MHW_{onset}	MHW_{end}	Long.°E	Lat.°N	Mean Duration (Days)	i_{max} (°C)	i_{mean} (°C)	i_{cum} (°C Days)
24-04-2013	08-05-2013	24.375	40.875	15	5.06	3.92	58.77
28-08-2020	31-12-2020	34.625	34.375	126	3.41	2.54	319.65
14-05-2020	20-05-2020	28.500	34.250	7	6.35	3.51	24.57

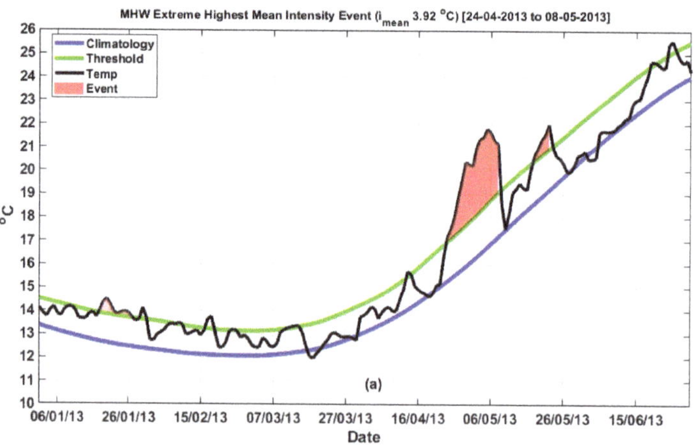
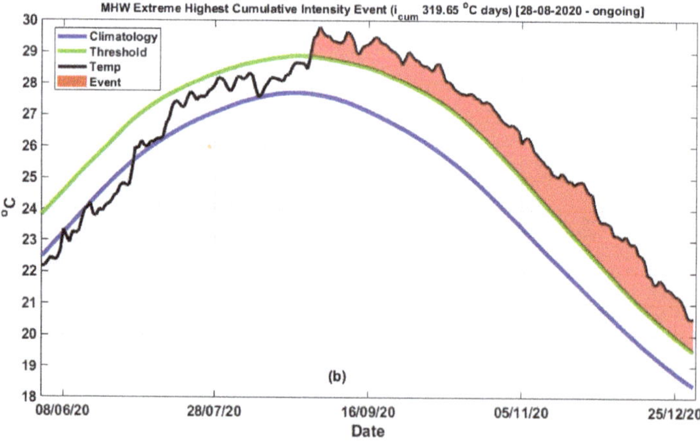

Figure 11. The EMED extreme MHW significant events over the study period (1982–2020), The SST climatology from NOAA OI SST for the detection of MHWs (blue), 90th percentile MHW threshold (green), and SST time series (black) for each MHW. The pink filled area indicates the period associated with the identified MHW, where (**a**) MHW extreme highest mean intensity event (i_{mean} 3.92 °C) (24-04-2013 to 08-05-2013), (**b**) MHW extreme highest cumulative intensity event (i_{cum} 319.65 °C days) (28-08-2020—ongoing).

The EMED Most Intense [14–20 May 2020] MHW Event over the Study Period (1982–2020)

In this subsection, the authors will describe in detail the most significant intense event of 14–20 May 2020. We found the most significant intense MHW event located at longitude ~28.500° E and latitude ~34.250° N south of Rhodes. Two more MHW significant events were discovered to be associated with the most intense main event, according to our analysis. The three significant events occur in the Levantine basin. It is critical in our research to assist in identifying and quantifying the main atmospheric causes of these extreme events. Table 3 demonstrates the EMED most intense [14–20 May 2020] MHW significant event characteristics, including duration, i_{max}, i_{mean}, i_{cum}, wind stress magnitude, 2m air temperature and mean sea level pressure. The i_{max} for the four events are 6.35, 5.66, 4.74, and 2.95 °C, respectively.

Table 3. The EMED most intense (14–20 May 2020) MHW events characteristics (mean duration (Days), i_{max} (°C), i_{mean} (°C), i_{cum} (°C days)), wind stress magnitude (N/m^2), 2 m air temperature (°C) and mean sea level pressure (mb) with the corresponding Longitude and Latitude in degrees.

Long.° E	Lat.° N	i_{max} (°C)	i_{mean} (°C)	i_{cum} (°C Days)	Wind Stress Magnitude (N/m^2)	2m Air Temp. (°C)	Msl Pressure (mb)
28.500	34.250	6.35	3.51	24.57	2.7×10^{-4}	27.71	1015.2
24.625	35.875	5.66	3.25	22.78	6.2×10^{-5}	27.69	1014.5
34.125	36.125	4.74	3.34	26.71	1.3×10^{-3}	26.06	1014.1

Figure 12a shows the SST anomaly event map created by subtracting the historical climatological mean value at each grid from the mean temperature 14–20 May 2020 at the same location (grid). The description of temporal evolution of the SST most intense MHW event based on threshold 90th percentile and SST climatology are shown in Figure 12b. The spatial distribution patterns of the wind stress vectors combined with magnitude, 2 m mean air temperature and mean sea level pressure of ERA5 were chosen at the same time and location as the SST anomaly analysis, as shown in Figure 12c–e. We can see an interesting relationship between the SST anomaly, wind stress, and 2 m mean air temperature fields. We observed a reduction in the magnitude of wind stress fields (less than 0.001 N/m^2) at the same locations as the most intense MHW events (see Figure 12c and Table 3). The comparison of wind stress fields and SST anomaly fields revealed that MHW extreme events are related to wind stress reduction, which is consistent with the findings of [14–16,18]. Another intriguing parallel is found between the SST anomaly and the 2 m mean air temperature patterns. High air temperature values (>25 °C) were found at the same locations as the MHW extreme events, as shown in Figure 12d and Table 3. Furthermore, we observed a very persistent ridge of high atmospheric pressure >1014 mb over the MHW extreme events locations in the EMED as shown in Figure 12e and Table 3. In conclusion, the atmospheric conditions, such as, mean sea level pressure, wind stress, and air temperature, may be linked to MHWs existence, Figure 12c–e.

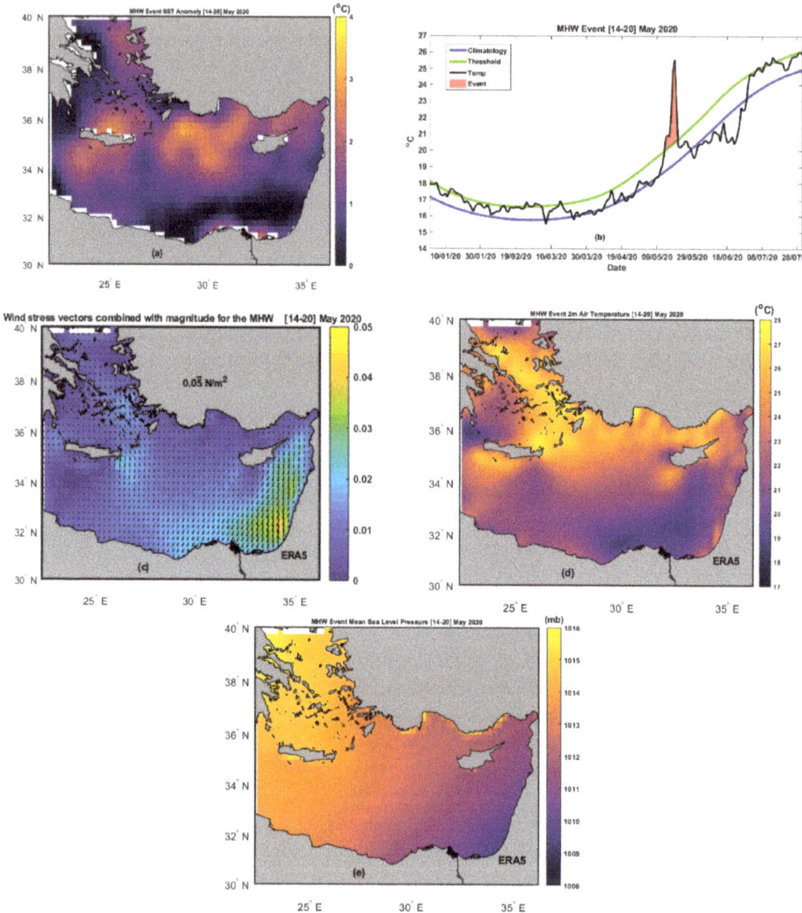

Figure 12. The EMED most significant intense (14–20 May 2020) MHW events over the study period (1982–2020), where, (**a**) MHW mean SST anomaly °C. (**b**) SST climatology from NOAA OI SST for the detection of MHWs (blue), 90th percentile MHW threshold (green), and SST time series (black) for each MHW. The pink filled area indicates the period associated with the identified MHW event. (**c**) Wind stress vectors combined with magnitude for the MHW in N/m^2 based on ERA5 hourly data. (**d**) 2 m mean air temperature °C based on ERA5 hourly data. (**e**) Mean sea level pressure (mb) based on ERA5 hourly data.

6. Discussion

According to our SST climatology analysis (Figures 2 and 3), the maximum values over the four decades were detected during the last decade, which agrees with [19,35,36]. Table 1 have showed that the last decade was the warmest decade by about 0.94 °C during the entire 1982–2020 study period in agreement with [51,52]. During the last decade, the SST anomaly patterns revealed a warming of about 0.10 to 0.50 °C over the northern EMED basin, with the exception of some parts of the north Ionian Sea (Figure 3d). The authors have attributed this to the effect of BiOS which is responsible for decadal reversals of the Ionian basin-wide circulation as mentioned by [37–41]. The maximum SST anomaly for the last decade was 0.50 °C in the northwest Aegean Sea, while the minimum value was −0.50 °C along EMED's southwestern coast (Figure 3d).

The authors have discovered a high spatial variability of SST linear trend over the EMED, ranging from 0.1 to 0.5 °C per decade (Figure 4). The basin average warming rate was about $0.33 \pm 0.04\,°C$ per decade. Over the last twenty years, several SST studies for the Mediterranean and the EMED region were published, addressing trends and changes in ocean surface temperature [6,19,28,35,52–58]. Ref. [53] have observed a trend of $0.3\,°C$ per decade in the western basin and $0.5\,°C$ per decade in the eastern basin over the period from 1985 to 2006. Similar results were recorded in [28], with an average mean warming rate of around $0.37\,°C$ per decade for the whole Mediterranean basin during 1985–2008, and $0.26\,°C$ per decade and $0.42\,°C$ per decade for the western and eastern sub-basins, respectively. Over a 25-year (1993–2017), the warming rate in the Mediterranean was about $0.36\,°C$ per decade, which was highly correlated with sea level trend, especially in the EMED [52]. According to [35], the annual warming trend for the deseasonalized Mediterranean global averaged SST was $0.36\,°C$ per decade over the period 1982–2016. More recently, studies with longer time series, such as [19], estimated a $0.35\,°C$ per decade warming for the global Mediterranean basin for the 38-year period 1982–2019. Figure 5 have revealed the highest negative peaks were observed in the winters of 1987 and 2019 in accordance with [46], The lower peak values (i.e., cold periods) in the 1990s could be attributed to the eruption of Mount Pinatubo in the Philippines, which caused a cold air-temperature anomaly throughout the Middle East during the winter of 1992, as stated by [47].

We have shown in Figure 6 the spatial distribution and trends per decade of the MHWs main characteristics (mean frequency, duration, (i_{cum}) and (i_{max})) overlaid with no significant values $(p > 0.05)$ for the entire study period. According to Figure 6c–f we found the patterns of i_{cum} (trend) and mean duration (trend) were nearly identical. The MHW mean duration has increased significantly $(p < 0.05)$ with values > 17 days with a trend of > 10 days per decade in front of the Egyptian coast in the southeast of the Levantine Sea and off the Israeli coast (Figure 6d). No significant values $(p > 0.05)$ were found in MHW duration at southeast of Crete, off Mersa Matruh city and the north Ionian Sea region. This could be attributed to the accuracy and precision of SST NOAA AVHRR data near coastlines, as described by [59,60]. According to their research, the accuracy and precision of Earth Observation (EO) SST data at the coastline are not well defined. According to Figure 7, the Levantine basin has more significant $(p < 0.05)$ MHW events (>4) in the last decade than the Ionian and Aegean Seas.

In the last decade, the highest significant MHW $(p < 0.05)$ duration > 27 days and i_{cum} values > 35 (°C days) were detected off the Israeli coast (Figures 8d and 9d), near the Shikmona anti-cyclonic gyre location [43,50]. The MHW i_{max} spatial distribution pattern were significant $(p < 0.05)$ throughout the entire Levantine basin over the last decade as demonstrated in Figure 10d. The significant MHW i_{max} values have ranged from $1.42\,°C$ off Mersa Matruh to more than $2.5\,°C$ in the northeast Aegean basin, as presented in Figure 10d. An extreme MHW event was discovered in 2013, with the maximum MHW i_{mean} temperature of $3.92\,°C$ and lasted 15 days (24 April–08 May) with an i_{max} of $5.06\,°C$ and i_{cum} of $58.77\,°C$ days (Figure 11a and Table 2). While, the most severe event (i_{cum} of $319.65\,°C$ days) was still ongoing at the time of the analysis with 126 days, i_{max} of $3.41\,°C$ and i_{mean} of $2.54\,°C$, Figure 11b. The most significant intense MHW event was located south of Rhodes at longitude $\sim28.500°$ E and latitude $\sim34.250°$ N and was associated with another two extreme events located in Levantine basin, that occurred simultaneously in 14–20 May 2020 with very high intensities (Figure 12a and Table 3). At the same locations as the most intense MHW events, we have observed a reduction in the magnitude of wind stress fields (less than $0.001\,N/m^2$) (see Figure 12c and Table 3). We have concluded that the MHW extreme events are related to the wind stress reduction in agreement with [14–16,18].

7. Conclusions

This study has quantified the SST interannual variability, trends, and spatial distribution of main MHW characteristics in the EMED over 39 years (1982–2020).

Some specific hotspot areas in the Mediterranean Sea as described by [61] such as Aegean, north Levantine and north Ionian have been recorded as the highest SST anomaly patterns during P_4 (2011–2020) which agrees with [18]. During last decade (P_4 (2011–2020)), the northern EMED basin's mean SST anomaly was nearly 1.00 °C higher than the southern one. The minimum SST anomaly value found -0.50 °C and was located on the southwestern coast of EMED in agreement with [18]. This could be due to upwelling associated with the Rhodes cyclonic gyre in the northern EMED basin, as mentioned in [42].

The SST EOF analysis revealed that the Ionian Sea had the highest interannual variability, while the Aegean had the lowest.

Over the entire study period, there was a clear decadal increase in MHW frequency in all EMED locations. The MHW frequency trend values were significantly ($p < 0.05$) positive. The correlation analysis between the mean MHWs spatial distribution of duration and frequency revealed a strong negative relationship, which was consistent with the findings of [3]. The authors found that the mean MHW frequency and duration increased by 40% and 15%, respectively

We have characterized trends and variability of MHWs for the EMED from 1982 to 2020 based on satellite NOAA OI SST V2.1 data and a unified MHW framework. We have shown that between 1982 and 2020, on EMED average, MHW frequency trend increased by 1.2 events per decade and the average MHW i_{cum} trend increased by 5.4 °C days.

The MHW main characteristics patterns over the last decade have shown greater significant values in most of the Levantine Sea. There were no significant ($p > 0.05$) found in the MHW main characteristics at southeast of Crete, off Mersa Matruh city and the north Ionian Sea region over the entire the study period.

The authors have showed the significant EMED extreme MHW events characteristics (mean duration, i_{max}, i_{mean}, and i_{cum}) over the study period (1982–2020). During these extreme events, there was a similarity in the SST anomaly, wind stress, and 2 m mean air temperature fields. We observed a reduction in the magnitude of wind stress fields (less than 0.001 N/m^2) at the same locations as the most intense MHW events. High air temperature values (>25 °C) have been linked to the locations of MHW extreme events. We concluded that atmospheric conditions such as mean sea level pressure, wind stress, and air temperature could be related to the existence of MHW.

In the future, we should focus more on the anthropogenic factors that contribute to the increasing severity and recurrence of MHW events. Furthermore, possible MHW links to different climatic and atmospheric indices [13,62–64] such as North Atlantic Oscillations (NAO), East Atlantic (EA), East Atlantic–West Russia (EAW), Mediterranean Oscillation (MO) and North-Sea Caspian Pattern (NCP). There is a lot of speculation about what caused the weather pattern we saw during this most severe MHW event. Was it just a coincidental occurrence? More information about the detection, characterization, impact assessment, and prediction of MHWs could be obtained.

Author Contributions: Conceptualization, O.I., B.M. and H.N.; methodology, H.N. and O.I.; formal analysis, O.I., H.N. and B.M., investigation, O.I., B.M. and H.N.; resources, O.I., B.M. and H.N.; data curation, H.N. and O.I.; writing—original draft preparation, O.I., B.M. and H.N.; writing—review and editing H.N., O.I. and B.M.; visualization, O.I., B.M. and H.N.; supervision, H.N. and O.I.; project administration, funding acquisition. This research received no external funding. All authors have read and agreed to the published version of the manuscript.

Funding: This research received no external funding.

Institutional Review Board Statement: Not applicable.

Informed Consent Statement: Not applicable.

Data Availability Statement: Not applicable.

Acknowledgments: The authors would like to thank Nadia Pinardi for her constructive remarks. The authors also would like to thank Khaled Elhusseiny, MSc Translation, York University and Doha Shahin, Language Interpreters Instructor at Conestoga College for their diligent proofreading of this article. Finally, we would like to express our sincere gratitude to Luke Marsden from UNIS in Norway for his wonderful collaboration in editing our manuscript. Special thanks to the three anonymous reviewers for their constructive comments, which helped to improve the manuscript.

Conflicts of Interest: No conflict of interest.

References

1. Frölicher, T.L.; Laufkötter, C. Emerging risks from marine heat waves. *Nat. Commun.* **2018**, *9*, 650. [CrossRef]
2. Manta, G.; de Mello, S.; Trinchin, R.; Badagian, J.; Barreiro, M. The 2017 record marine heatwave in the Southwestern Atlantic shelf. *Geophys. Res. Lett.* **2018**, *45*. [CrossRef]
3. Oliver, E.C.J.; Donat, M.G.; Burrows, M.T.; Moore, P.J.; Smale, D.A.; Alexander, L.V.; Benthuysen, J.A.; Feng, M.; Gupta, A.S.; Hobday, A.J.; et al. Longer and more frequent marine heatwaves over the past century. *Nat. Commun.* **2018**, *9*, 1324. [CrossRef]
4. Hobday, A.J.; Alexander, L.V.; Perkins, S.E.; Smale, D.A.; Straub, S.C.; Oliver, E.C.; Benthuysen, J.A.; Burrows, M.T.; Donat, M.G.; Feng, M.; et al. A hierarchical approach to defining marine heatwaves. *Prog. Oceanogr.* **2016**, *141*, 227–238. [CrossRef]
5. Hobday, A.J.; Oliver, E.C.; Gupta, A.S.; Benthuysen, J.A.; Burrows, M.T.; Donat, M.G.; Holbrook, N.J.; Moore, P.J.; Thomsen, M.S.; Wernberg, T.; et al. Categorizing and naming marine heatwaves. *Oceanography* **2018**, *31*, 162–173. [CrossRef]
6. Strub, P.T.; James, C.; Montecino, V.; Rutllant, J.A.; Blanco, J.L. Ocean circulation along the southern Chile transition region (38°–46°S): Mean, seasonal and interannual variability, with a focus on 2014–2016. *Prog. Oceanogr.* **2019**, *172*, 159–198. [CrossRef]
7. Holbrook, N.J.; Scannell, H.A.; Gupta, A.S.; Benthuysen, J.A.; Feng, M.; Oliver, E.C.J.; Alexander, L.V.; Burrows, M.T.; Donat, M.G.; Hobday, A.J.; et al. A global assessment of marine heatwaves and their drivers. *Nat. Commun.* **2019**, *10*, 2624. [CrossRef] [PubMed]
8. Jacox, M.G. Marine heatwaves in a changing climate. *Nature* **2019**, *571*, 485–487. [CrossRef]
9. Garrabou, J.; Coma, R.; Bensoussan, N.; Bally, M.; Chevaldonné, P.; Cigliano, M.; Diaz, D.; Harmelin, J.G.; Gambi, M.; Kersting, D. Mass mortality in Northwestern Mediterranean rocky benthic communities: Effects of the 2003 heat wave. *Glob. Chang. Biol.* **2009**, *15*, 1090–1103. [CrossRef]
10. Marba, N.; Duarte, C.M. Mediterranean warming triggers seagrass (posidonia oceanica) shoot mortality. *Glob. Chang. Biol.* **2010**, *16*, 2366–2375. [CrossRef]
11. Malanotte-Rizzoli, P. Introduction to special section: Physical and biochemical evolution of the eastern Mediterranean in the 1990s (PBE). *J. Geophys. Res.* **2003**, *108*, 8100. [CrossRef]
12. Borzelli, G.L.E.; Gacic, M.; Cardin, V.; Civitarese, G. Eastern Mediterranean Transient and Reversal of the Ionian Sea Circulation. *Geophys. Res. Lett.* **2009**, *36*, L15108. [CrossRef]
13. Pinardi, N.; Cessi, P.; Borile, F.; Wolfe, C.L.P. The Mediterranean Sea overturning circulation. *J. Phys. Oceanogr.* **2019**, *49*, 1699–1721. [CrossRef]
14. Grazzini, F.; Viterbo, P. Record-breaking warm sea surface temperature of the Mediterranean Sea. *ECMWF Newslett.* **2003**, *98*, 30–31.
15. Sparnocchia, S.; Schiano, M.; Picco, P.; Bozzano, R.; Cappelletti, A. The anomalous warming of summer 2003 in the surface layer of the Central Ligurian Sea (Western Mediterranean). *Ann. Geophys.* **2006**, *24*, 443–452. [CrossRef]
16. Olita, A.; Sorgente, R.; Natale, S.; Ribotti, A.; Bonanno, A.; Patti, B. Effects of the 2003 European heatwave on the Central Mediterranean Sea: Surface fluxes and the dynamical response. *Ocean Sci.* **2007**, *3*, 273–289. [CrossRef]
17. Mavrakis, A.F.; Tsiros, I.X. The abrupt increase in the Aegean Sea surface temperature during the June 2007 southeast Mediterranean heatwave-a marine heatwave event? *Weather* **2018**, *74*, 201–207. [CrossRef]
18. Darmaraki, S.; Somot, S.; Sevault, F.; Nabat, P.; Narvaez, W.D.C.; Cavicchia, L.; Djurdjevic, V.; Li, L.; Sannino, G.; Sein, D.V. Future evolution of Marine Heatwaves in the Mediterranean Sea. *Clim. Dyn.* **2019**, *53*, 1371–1392. [CrossRef]
19. Pastor, F.; Valiente, J.A.; Khodayar, S. A Warming Mediterranean: 38 Years of Increasing Sea Surface Temperature. *Remote Sens.* **2020**, *12*, 2687. [CrossRef]
20. Pinardi, N.; Zavatarelli, M.; Arneri, E.; Crise, A.; Ravaioli, M. The physical, sedimentary and ecological structure and variability of shelf areas in the Mediterranean Sea. In *The Sea*; Robinson, A.R., Brink, K.H., Eds.; Harvard University Press: Cambridge, MA. USA, 2005; Volume 14B, Chapter 32; pp. 1245–1331.
21. Nagy, H.; Elgindy, A.; Pinardi, N.; Zavatarelli, M.; Oddo, P. A nested pre-operational model for the Egyptian shelf zone: Model configuration and validation/calibration. *J. Dyn. Atmos. Ocean.* **2017**, *80*, 75–96. [CrossRef]
22. Reynolds, R.W.; Smith, T.M.; Liu, C.; Chelton, D.B.; Casey, K.S.; Schlax, M.G. Daily high-resolution-blended analyses for sea surface temperature. *J. Clim.* **2007**, *20*, 5473–5496. [CrossRef]
23. Thomson, R.E.; Emery, W.J. *Data Analysis Methods in Physical Oceanography*, 3rd ed.; Elsevier Inc.: Amsterdam, The Netherlands, 2014; ISBN 9780123877833.
24. Arafeh-Dalmau, N.; Montaño-Moctezuma, G.; Martínez, J.A.; Beas-Luna, R.; Schoeman, D.S.; Torres-Moye, G. Extreme Marine Heatwaves Alter Kelp Forest Community Near Its Equatorward Distribution Limit. *Front. Mar. Sci.* **2019**, *6*, 499. [CrossRef]
25. Wilks, D.S. *Statistical Methods in the Atmospheric Sciences*; Academic Press: Cambridge, MA, USA, 2011; ISBN 0123850223.

26. Hamed, K.H.; Ramachandra Rao, A. A modified Mann-Kendall trend test for autocorrelated data. *J. Hydrol.* **1998**, *204*, 182–196. [CrossRef]
27. Wang, F.; Shao, W.; Yu, H.; Kan, G.; He, X.; Zhang, D.; Ren, M.; Wang, G. Re-evaluation of the Power of the Mann-Kendall Test for Detecting Monotonic Trends in Hydrometeorological Time Series. *Front. Earth Sci.* **2020**, *8*, 14. [CrossRef]
28. Skliris, N.; Sofianos, S.; Gkanasos, A.; Mantziafou, A.; Vervatis, V.; Axaopoulos, P.; Lascaratos, A. Decadal scale variability of sea surface temperature in the Mediterranean Sea in relation to atmospheric variability. *Ocean Dyn.* **2012**, *62*, 13–30. [CrossRef]
29. Trenberth, K.E. Signal versus noise in the Southern Oscillation. Monthly. *Weather Rev.* **1984**, *112*, 326–333. [CrossRef]
30. Carlson, D.F.; Yarbo, L.A.; Scolaro, S.; Poniatowski, M.; Absten, V.M.; Carlson, P.R., Jr. Sea surface temperatures and seagrass mortality in Florida Bay: Spatial andtemporal patterns discerned from MODIS and AVHRR data. *Remote Sens. Environ.* **2018**, *208*, 171–188. [CrossRef]
31. Zhao, Z.; Marin, M. A MATLAB toolbox to detect and analyze marine heatwaves. *J. Open Source Softw.* **2019**, *4*, 1124. [CrossRef]
32. Hersbach, H.; Bell, B.; Berrisford, P.; Hirahara, S.; Horányi, A.; Muñoz-Sabater, J.; Nicolas, J.; Peubey, C.; Radu, R.; Schepers, D.; et al. The ERA5 global reanalysis. *Q. J. R. Meteorol. Soc.* **2020**, *146*, 1999–2049. [CrossRef]
33. Nagy, H.; Lyons, K.; Nolan, G.; Cure, M.; Dabrowski, T. A Regional Operational Model for the North East Atlantic: Model Configuration and Validation. *J. Mar. Sci. Eng.* **2020**, *8*, 673. [CrossRef]
34. Hellerman, S.; Rosenstein, M. Normal monthly wind stress over the world ocean with error estimates. *Phys. Oceanogr.* **1983**, *13*, 1093–1104. [CrossRef]
35. Pastor, F.; Valiente, J.A.; Palau, J.L. Sea surface temperature in the Mediterranean: Trends and spatial patterns (1982–2016). *Pure Appl. Geophys.* **2018**, *175*, 4017–4029. [CrossRef]
36. El-Geziry, T.M. Long-term changes in sea surface temperature (SST) within the southern Levantine Basin. *Acta Oceanol. Sin.* **2021**, *40*, 27–33. [CrossRef]
37. Gacic, M.; Civitarese, G.; EusebiBorzelli, G.L.; Kovacevic, V.; Poulain, P.M.; Theocharis, A.; Menna, M.; Catucci, A.; Zarokanellos, N. On the relationship between the decadal oscil-lations of the Northern Ionian Sea and the salinity distributions in the Eastern Mediterranean. *J. Geophys. Res.* **2011**, *116*, C12002. [CrossRef]
38. Gacic, M.; Civitarese, G.; Kovacevic, V.; Ursella, L.; Bensi, M.; Menna, M.; Cardin, V.; Poulain, P.M.; Cosoli, S.; Notarstefano, G.; et al. Extreme winter 2012 in the Adriatic: An example of climatic effect on the BiOS rhythm. *Ocean Sci.* **2014**, *10*, 513–522. [CrossRef]
39. Mkhinini, N.; Coimbra, A.L.S.; Stegner, A.; Arsouze, T.; Taupier-Letage, I.; Béranger, K. Long-lived mesoscale eddies in the eastern Mediterranean Sea: Analysis of 20 years of AVISO geostrophic velocities. *J. Geophys. Res. Ocean.* **2014**, *119*, 8603–8626. [CrossRef]
40. Nagy, H.; Di-Lorenzo, E.; El-Gindy, E. The Impact of Climate Change on Circulation Patterns in the Eastern Mediterranean Sea Upper Layer Using Med-ROMS Model. *Prog. Oceanogr.* **2019**, *175*, 244. [CrossRef]
41. Rubino, A.; Gačić, M.; Bensi, M.; Kovačević, V.; Malačič, V.; Menna, M.; Negretti, M.E.; Sommeria, J.; Zanchettin, D.; Barreto, R.V.; et al. Experimental evidence of long-term oceanic circulation reversals without wind influence in the North Ionian Sea. *Sci. Rep.* **2020**, *10*, 1905. [CrossRef]
42. Pinardi, N.; Masetti, E. Variability of the large-scale General Circulation of the Mediterranean Sea from observation and modeling: A review. *Palaeogeogr. Palaeoclimatol. Palaeoecol.* **2000**, *158*, 153–173. [CrossRef]
43. Amitai, Y.; Lehahn, Y.; Lazar, A.; Heifetz, E. Surface circulation of the eastern Mediterranean Levantine basin: Insights from analyzing 14 years of satellite altimetry data. *J. Geophys. Res.* **2010**, *115*, C10058. [CrossRef]
44. Carniel, S.; Bonaldo, D.; Benetazzo, A.; Bergamasco, A.; Boldrin, A.; Falcieri, F.M.; Sclavo, M.; Trincardi, F.; Falcieri, F.M.; Langone, L.; et al. Off-Shelf Fluxes across the Southern Adriatic Margin: Factors Controlling Dense Water-Driven Transport Phenomena. *Mar. Geol.* **2016**, *375*, 44–63. [CrossRef]
45. Chhak, K.; Di Lorenzo, E. Decadal variations in the California Current upwelling cells. *Geophys. Res. Lett.* **2007**, *34*. [CrossRef]
46. Tolika, K.; Maheras, P.; Pytharoulis, I.; Anagnostopoulou, C. The anomalous low and high temperatures of 2012 over Greece—An explanation from a meteorological and climatological perspective. *Nat. Hazards Earth Syst. Sci.* **2014**, *14*, 501–507. [CrossRef]
47. Genin, A.; Lazar, B.; Brenner, S. Vertical mixing and coral death in the Red Sea following the eruption of Mount Pinatubo. *Nature* **1995**, *377*, 507–510. [CrossRef]
48. Galli, G.; Solidoro, C.; Lovato, T. Marine heat waves hazard 3D maps, and the risk for low motility organisms in a warming Mediterranean Sea. *Front. Mar. Sci.* **2017**, *4*, 136. [CrossRef]
49. Brenner, S. Structure, and evolution of warm core eddies in the Eastern Mediterranean Levantine Basin. *J. Geophys. Res.* **1989**, *94*, 12593–12602. [CrossRef]
50. Brenner, S. Long term evolution and dynamics of a persistent warm Core eddies in the Eastern Mediterranean Sea. *J. Deep Sea Res.* **1993**, *40*, 1193–1206. [CrossRef]
51. Minnett, P.J.; Azcárate, A.A.; Chin, T.M.; Corlett, G.K.; Gentemann, C.L.; Karagali, I.; Li, X.; Marsouin, A.; Marullo, S.; Maturi, E.; et al. Half a century of satellite remote sensing of sea-surface temperature. *Remote Sens. Environ.* **2019**, *233*, 111366. [CrossRef]
52. Mohamed, B.; Abdallah, A.M.; Din, K.A.E.; Nagy, H.; Shaltout, M.A. Inter-annual variability and trends of sea level and sea surface temperature in the mediterranean sea over the last 25 years. *Pure Appl. Geophys.* **2019**, *176*, 3787–3810. [CrossRef]
53. Nykjaer, L. Mediterranean sea surface warming 1985–2006. *Clim. Res.* **2009**, *39*, 11–17. [CrossRef]

54. Rayner, N.A.; Brohan, P.; Parker, D.E.; Folland, C.K.; Kenndy, J.J.; Vanicek, M.; Ansell, T.J.; Tett, S.F.B. Improved analysis of changes and uncertainties in sea surface temperature measured in situ since the mid-ninteenth century: The HadSST2 dataset. *J. Clim.* **2006**, *19*, 446–469. [CrossRef]
55. Marullo, S.; Artale, V.; Santoleri, R. The SST multidecadal variability in the Atlantic-Mediterranean region and its relation to AMO. *J. Clim.* **2011**, *24*, 4385–4401. [CrossRef]
56. Banzon, V.; Smith, T.M.; Chin, T.M.; Liu, C.; Hankins, W. A long-term record of blended satellite and in situ sea-surface temperature for climate monitoring, modeling, and environmental studies. *Earth Syst. Sci. Data* **2016**, *8*, 165–176. [CrossRef]
57. Fiedler, E.K.; McLaren, A.; Banzon, V.; Brasnett, B.; Ishizaki, S.; Kennedy, J.; Rayner, N.; Jones, J.R.; Corlett, G.; Merchant, C.J.; et al. Intercomparison of long-term sea surface temperature analyses using the GHRSST Multi-Product Ensemble (GMPE) system. *Remote Sens. Environ.* **2019**, *222*, 18–33. [CrossRef]
58. López García, M.J. SST Comparison of AVHRR and MODIS Time Series in the Western Mediterranean Sea. *Remote Sens.* **2020**, *12*, 2241. [CrossRef]
59. Brewin, R.J.W.; de Mora, L.; Billson, O.; Jackson, T.; Russell, P.; Brewin, T.G.; Shutler, J.; Miller, P.I.; Taylor, B.H.; Smyth, T.; et al. Evaluating operational AVHRR sea surface temperature data at the coastline using surfers. *Estuar. Coast. Shelf Sci.* **2017**, *196*, 276–289. [CrossRef]
60. Brewin, R.J.W.; Smale, D.A.; Moore, P.J.; Dall'Olmo, G.; Miller, P.I.; Taylor, B.H.; Smyth, T.J.; Fishwick, J.R.; Yang, M. Evaluating Operational AVHRR Sea Surface Temperature Data at the Coastline Using Benthic Temperature Loggers. *Remote Sens.* **2018**, *10*, 925. [CrossRef]
61. Giorgi, F. Climate change hot-spots. *Geophys. Res. Lett.* **2006**, *33*. [CrossRef]
62. Kutiel, H.; Benaroch, Y. North Sea-Caspian Pattern (NCP)—An upper level atmospheric teleconnection aecting the Eastern Mediterranean: Identication and definition. *Theor. Appl. Climatol.* **2002**, *71*, 17–28. [CrossRef]
63. Brunetti, M.; Kutiel, H. The relevance of the North-Sea Caspian Pattern (NCP) in explaining temperature variability in Europe and the Mediterranean. *Nat. Hazards Earth Syst. Sci.* **2011**, *11*, 2881–2888. [CrossRef]
64. Criado-Aldeanueva, F.; Soto-Navarro, J. Climatic Indices over the Mediterranean Sea: A Review. *Appl. Sci.* **2020**, *10*, 5790. [CrossRef]

Journal of
*Marine Science
and Engineering*

MDPI

Article

CMIP5-Based Projection of Decadal and Seasonal Sea Surface Temperature Variations in East China Shelf Seas

Huiqiang Lu [1], Chuan Xie [2], Cuicui Zhang [1,*] and Jingsheng Zhai [1,*]

[1] School of Marine Science and Technology, Tianjin University, 92 WeiJin RD, Nankai District, Tianjin 300072, China; chentu90@163.com
[2] School of Oceanography, Shanghai Jiaotong University, 800 DongChuan RD, Minhang District, Shanghai 200240, China; xiechuan@sjtu.edu.cn
* Correspondence: cuicui.zhang@tju.edu.cn (C.Z.); jingsheng@tju.edu.cn (J.Z.)

Abstract: The East China Shelf Seas, comprising the Bohai Sea, the Yellow Sea, and the shelf region of East China Sea, play significant roles among the shelf seas of the Western North Pacific Ocean. The projection of sea surface temperature (SST) changes in these regions is a hot research topic in marine science. However, this is a very difficult task due to the lack of available long-term projection data. Recently, with the high development of simulation technology based on numerical models, the model intercomparison projects, e.g., Phase 5 of the Climate Model Intercomparison Project (CMIP5), have become important ways of understanding climate changes. CMIP5 provides multiple models that can be used to estimate SST changes by 2100 under different representative concentration pathways (RCPs). This paper developed a CMIP5-based SST investigation framework for the projection of decadal and seasonal variation of SST in East China Shelf Seas by 2100. Since the simulation results of CMIP5 models may have degrees of errors, this paper uses hydrological observation data from World Ocean Atlas 2018 (WOA18) for model validation and correction. This paper selects seven representative ones including ACCESS1.3, CCSM4, FIO-ESM, CESM1-CAM5, CMCC-CMS, NorESM1-ME, and Max Planck Institute Earth System Model of medium resolution (MPI-ESM-MR). The decadal and seasonal SST changes in the next 100 years (2030, 2060, 2090) are investigated by comparing with the present analysis in 2010. The experimental results demonstrate that SST will increase significantly by 2100: the decadal SST will increase by about 1.55 °C, while the seasonal SST will increase by 1.03–1.95 °C.

Keywords: decadal and seasonal SST variation; East China Shelf Seas; CMIP5; WOA18

Citation: Lu, H.; Xie, C.; Zhang, C.; Zhai, J. CMIP5-Based Projection of Decadal and Seasonal Sea Surface Temperature Variations in East China Shelf Seas. *J. Mar. Sci. Eng.* **2021**, *9*, 367. https://doi.org/10.3390/jmse9040367

Academic Editor: Kyung-Ae Park

Received: 31 January 2021
Accepted: 23 March 2021
Published: 30 March 2021

Publisher's Note: MDPI stays neutral with regard to jurisdictional claims in published maps and institutional affiliations.

1. Introduction

The climate is changing. Our Earth is warming up. Many agree that climate change may be one of the greatest threats facing our planet. Ocean warming, which contributes much to global warming, has become a more serious problem. Recent observation-based estimates and model simulation results show that ocean warming is accelerating and will continue in the next one hundred years [1]. Since global warming may lead to many ecological problems, more efforts need to be made in the assessment and projection of the warming rate of the oceans in the future. The sea surface temperature (SST), which is an important physical parameter of oceans, can reflect the effect of climate change. The estimation of SST variation has become a hot research topic in marine science.

The East China Shelf Seas, which consist of the Bohai Sea, the Yellow Sea, and the shelf region of East China Sea, have been recognized as the most significant marginal seas in the Western North Pacific Ocean (WNPO) [2]. The Bohai Sea is a shallow semi-enclosed sea with an average depth of only 18 m, and the changes of its water temperature have a greater influence on its ecosystem than the physical forcing from the external oceans [3]. The Yellow Sea is a wide and shallow sea, where the depth in most regions is less than 50 m. Its water column can be seriously affected by the atmospheric conditions such as heating,

163

cooling, and wind stress within it than from the open seas [4]. The East China Sea, which is located between the largest continent and the largest ocean, has been largely effected by the climatic forcing from both the high-latitude Northern Hemisphere (East Asian Monsoon System) and the tropic ocean (Kuroshio Current (KC)) and generates characteristic climate patterns with strong horizontal and vertical temperature gradients [5–7]. These East China Shelf Seas are very sensitive to global warming because of their shallow waters [8]. The SST increase over these regions is about 0.8 to 2 °C per century, which is nearly twice the global average increase of SST [9]. The SST increases can not only affect the metabolic rates of marine organisms, but also influence other oceanic states, such as local currents [3]. Therefore, we need to analyze the SST variation, especially to project the long-term spatial and temporal variations in the future. However, this is a very difficult task because of the lack of effective long-term projection data [8].

Recently, with the rapid development of ocean simulations, model intercomparison projects have become important ways to investigate climate changes. Among them, the Coupled Climate Model Intercomparison Project plays an important role [10]. Phase 5 of the Climate Model Intercomparison Project (CMIP5) is an international collaboration framework, which provides a multimodel context to help understand the responses of climate models to a common forcing with the aim of promoting the climate model projection and assessment for the Fifth Assessment Report (AR5) of the Intergovernmental Panel on Climate Change (IPCC) [11]. CMIP5 contains multiple models, which can be used to project climate changes and sea level rise in the future under different climate change scenarios or representative concentration pathways (RCPs) including RCP2.7, RCP4.5, RCP6.0, and RCP8.5. These scenarios correspond to the peak of the atmosphere radiative imbalance of 2.6, 4.5, 6.0, and 8.5 W/m^2 by 2100, respectively [12]. Following the previous CMIPs, CMIP5 is a new experimental framework, which can be widely applied for the analysis of decadal and seasonal climate changes in the future.

In the literature, several CMIP5-based SST projection works have been developed. For example, Zhou and Ying [13] analyzed the interannual variability of the SST over the Pacific in the historical simulation and future analysis under RCP4.5 and RCP8.5. Qu and Huang [14] investigated the decadal variability of the tropical Indian Ocean SST–South Asian High (SAH) relation, as well as its response to global warming. Qin and Xie [15] studied the connections between the precipitation extremes during 1953–2002 in the dry and wet regions of China and the SST in the eastern tropical Pacific Ocean based on two sets of observation data, 17 CMIP5 models, and nine regional climate models. Tachibana et al. [16] examined the western Indian Ocean SST biases among the CMIP5 models and found that the multimodel ensemble mean SST biases over the western equatorial Indian Ocean are warmer than the observations during the summer monsoon season. Song et al. [17] evaluated 18 CMIP5 models according to their capability of simulating the SST annual cycle in the eastern equatorial Pacific. Xu et al. [18] tested a number of previously proposed mechanisms responsible for the southeastern tropical Atlantic SST bias based on CMIP5 models. Zhao and Zhang [19] analyzed the impacts of SST warming in the tropical Indian Ocean on the projected change in summer rainfall over Central Asia based on historical and RCP8.5 experiments. Kucharski and Joshi [20] evaluated the teleconnection from the tropical South Atlantic SST anomalies to the Indian monsoon based on observations and CMIP5 model data. Levine et al. [21] examined the extent and impact of cold SST biases developing in the northern Arabian Sea in the CMIP5 multi-model ensemble. Langehaug et al. [22] investigated the projection of SST in the Nordic Seas and Barents Sea using initialized hindcast simulations performed with three climate models (MPI-ESM-LR, CNRM-CM5, IPSL-CM5-LR). For the Chinese coastal seas, a few works have also been presented to study the SST changes using CMIP5 models. For example, Huang et al. [23] evaluated the capacities of 17 selected CMIP5 models on the historical SST simulation in the South China Sea and projected the SST changes in the 21st Century under RCP2.6, RCP 4.5, and RCP8.5, respectively. Tan et al. [24] evaluated the variation trend of SST over offshore China in the 21st Century based on the selected CMIP5 models under RCP4.5. Song et al. [25] assessed the monthly, seasonal, and interannual SSTs in the China seas over 1960–2002 using five representative CMIP5 models.

Although existing works have made some achievements in the study of the SST changes in the East China Shelf Seas, there are still some problems unsolved. Firstly, existing works mainly focused on the study of annual and decadal variations of the SST in the East China Shelf Seas without investigating the seasonal variations, which may affect ocean organisms and the ecological environment more seriously (e.g., the changes of species distribution and the move up of the phenophase). Secondly, existing works lack precise ocean observation data for model validation. Existing works mainly used the low-resolution observation data HadISST from the Hadley Center in the U.K. ($1° \times 1°$). It is not accurate enough for the evaluation of high-resolution CMIP5 models (e.g., the Max Planck Institute Earth System Model of medium resolution (MPI-ESM-MR) of $0.1° \times 0.1°$). Moreover, existing works only used the observation data for model validation without establishing model correction or simulation result modification, which is a fundamental procedure to make projections more accurate. Thirdly, the models used in existing works are not very appropriate. Several high-resolution models (e.g., the MPI-ESM-MR) were not fully evolved in the existing works. Fourthly, the study region of existing works was very rough. Existing works mainly referred to the Chinese coastal seas as the rectangle region of $0° \text{ N}–45° \text{ N}$, $100° \text{ E}–140° \text{ E}$, which is a very coarse region to study on the SST changes in the East China Shelf Seas.

To solve these problems, this paper establishes an SST analysis framework, which can be used for the projection of both decadal and seasonal SST variation in the East China Shelf Seas using high-resolution CMIP5 data. This paper investigates both the decadal and seasonal SST variation. For the second problem, we introduce the high-resolution hydrological observation data WOA18 ($0.25° \times 0.25°$) from World Ocean Atlas 2018 [26] for model validation and calculate the error map of each of the CMIP5 model data for the simulation result modification. This strategy ensures the high accuracy of CMIP5 model data. To solve the third problem, we select the best model MPI-ESM-MR with the highest resolution ($0.1° \times 0.1°$) from seven representative models (ACCESS1.3, CCSM4, FIO-ESM, CESM1-CAM5, CMCC-CMS, NorESM1-ME, and MPI-ESM-MR). MPI-ESM-MR not only has low errors, but its high resolution also guarantees the specific analysis of local climate changes. To solve the fourth problem, we utilize the bathymetric depth map with a finer grid to define the study area (as shown in Figure 1). In this paper, the decadal and seasonal SST variation analysis is performed by comparing the SST simulation result on 2030, 2060, and 2090 with the present analysis in 2010 under RCP4.5. We use RCP4.5 as a representative scenario because it is a medium-mitigation emission scenario that stabilizes direct radiative forcing at 4.5 W/m^2 (650 ppm CO_2 equivalent) by 2100 [27,28]. Compared to RCP4.5, RCP2.6 is a mitigation scenario leading to a very low forcing level (at 2.6 W/m^2 by 2100), while RCP8.5 is a scenario with very high greenhouse gas emissions (at 8.5 W/m^2 by 2100). RCP4.5 is neither too high, nor too low for projection. To the best of our knowledge, this is the first work to investigate both decadal and seasonal SST variations in the East China Shelf Seas by 2100 using high-resolution CMIP5 data.

This paper is organized as follows: Section 2 introduces the data and the study area. The model validation and simulation result modification are performed by comparing the CMIP5 model data with WOA18 data in Section 3. Sections 4 and 5 analyze the projection results of decadal and seasonal SST variations, respectively. Section 6 concludes this paper.

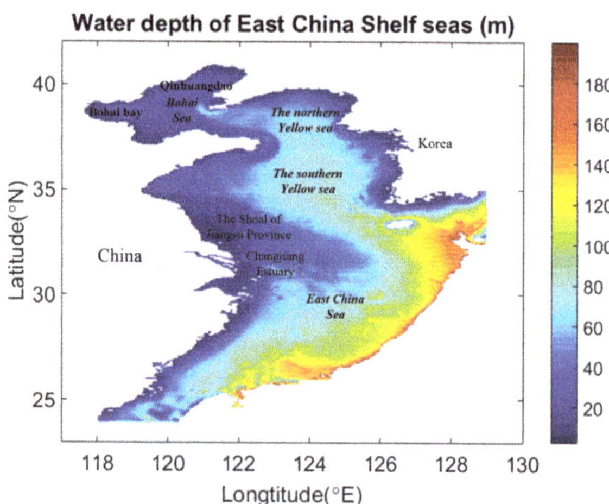

Figure 1. Depth map of East China Shelf Seas.

2. Data and Study Region

In this paper, CMIP5 model data were used for the projection of future SST changes, while WOA18 data were used for model validation, as well as for present analysis. CMIP5 consists of 36 models, which are developed by 16 institutes. In the preprocessing procedure, we selected the most representative models according to the following criteria: (1) for each institute, we selected the latest developed, with the highest resolution model for study; (2) the selected models should provide the simulation results of the SST from 2010 to 2100 continuously. We finally chose seven representative models. Their information is shown in Table 1. From Table 1, we can see that the resolution of Max Planck Institute Earth System Model of medium resolution (MPI-ESM-MR) is much higher than that of other models. MPI-ESM-MR couples general circulation models for the ocean and the atmosphere. It has been widely applied in many climate change experiments for either idealized CO_2-only forcing or forcings based on observations and RCP scenarios [29].

Table 1. CMIP5 models used in this paper: model name, average horizontal resolution (latitude × longitude), and reference. MPI-ESM-MR, Max Planck Institute Earth System Model of medium resolution.

Model Name	Horizontal Resolution	Reference
ACCESS1.3	0.3° × 1.0°	Bi et al., 2013 [30]
CCSM4	0.5° × 1.1°	Danabasoglu et al. 2012 [31]
FIO-ESM	0.5° × 1.1°	Qiao et al. 2013 [32]
CESM1-CAM5	0.9° × 1.25°	Meehl et al. 2013 [33]
CMCC-CMS	0.5° × 2.0°	Borrelli et al. 2012 [34]
NorESM1-ME	0.5° × 1.1°	Schwinger et al. 2016 [35]
MPI-ESM-MR	0.1° × 0.1°	Jungclaus et al. 2013 [36]

The annual and seasonal SST data in 2010 were obtained from the World Ocean Atlas 2018 (WOA18) hydrological observation averaged over 2005 to 2017. WOA18 provides both objectively analyzed (1° grid) climatological fields of in situ temperature for annual and seasonal compositing periods of the World Ocean. It also includes associated statistical observation data interpolated on 5°, 1°, and 0.25° grids [26]. We used the highest resolution data of 0.25° grid. As shown in Figure 1, the study area is comprised of the Bohai Sea, the Yellow Sea, and the shelf region of East China Sea.

Since each selected CMIP5 model has its inner simulation variations and errors, we calculated the climatological annual and seasonal water temperature in 2010 by the average result over 2006 to 2015. A similar processing was done for 2030, 2060, and 2090. They were investigated by the average result of 2026~2035, 2056~2064, and 2086~2094, respectively.

3. Model Validation and Simulation Result Modification

As we know, the simulation results of numeric ocean models may contain some degree of errors. Before using the CMIP5 model data, we needed to perform model validation to evaluate whether the simulation results are accurate enough for SST projection. The model validation was performed by comparing the annual and seasonal CMIP5 model data with the hydrological observation data of WOA18 in 2010. The qualitative and quantitative comparison results are shown in Figure 2 and Table 2, respectively. From Figure 2, we can find that most selected CMIP5 models have a similar SST distribution as the WOA18 data. However, different models also vary from each other in the model resolution and simulation accuracy. Among these models, the simulation result of MPI-ESM-MR is better than the others. Its resolution (in 0.1°) is the highest one (even higher than WOA18), and its simulation result is most similar to WOA18. MPI-ESM-MR can illustrate the local climate impact factors such as the Yellow Sea cold water mass in Summer and Autumn, the Yellow Sea warm current in Winter, the Kuroshio invasion, and the northern flow of the Taiwan warm current around Changjiang shore to the Tsushima Strait. Benefiting from the high resolution, its simulation result is even better than the observation data (WOA18). The numerical results in Table 2 also verify this finding. From Table 2, we can see that both the annual and seasonal simulation errors of MPI-ESM-MR are low enough. That means that MPI-ESM-MR is the best model that can be used for the projection of decadal and seasonal SST variations.

Table 2. Comparison of CMIP5 model data with the real observation data from WOA18 in 2010.

Data Source		Annual	Spring	Summer	Autumn	Winter
WOA18		18.07	16.80	25.79	18.49	11.20
	ACCESS1.3	0.67	1.49	2.42	0.61	0.62
	CCSM4	0.33	0.77	0.14	1.51	0.72
	FIO-ESM	0.79	0.17	-0.12	1.67	1.44
CMIP5 Model Errors	CESM1-CAM5	0.28	0.85	0.64	1.55	1.05
	CMCC-CMS	0.40	−0.50	−0.66	1.61	1.17
	NorESM1-ME	0.52	−0.40	−0.05	1.80	0.75
	MPI-ESM-MR	0.15	−0.96	0.81	1.67	0.72
	Average	0.26	−0.71	−0.70	1.52	0.94

After model validation, we performed simulation result modification to make the CMIP5 model data more accurate. Firstly, we calculated the error map of MPI-ESM-MR by comparing its results with WOA18 data for the year and the four seasons of 2010, respectively. Then, we subtracted these errors from the original simulation results to obtain more accurate projection data. Figure 3 illustrates the annual and seasonal error maps of MPI-ESM-MR. From Figure 3a–e, we can see that the error of MPI-ESM-MR is mainly concentrated in the deep trough of the Yellow Sea, where its simulation on the influence of the Yellow Sea warm current is stronger than the observation result. This phenomenon leads to the higher SST simulation in the eastern Yellow Sea in spring and winter and in the northern part of East China Shelf Seas in autumn than the WOA18 data. However, its simulation results of other regions were satisfactory enough for projection.

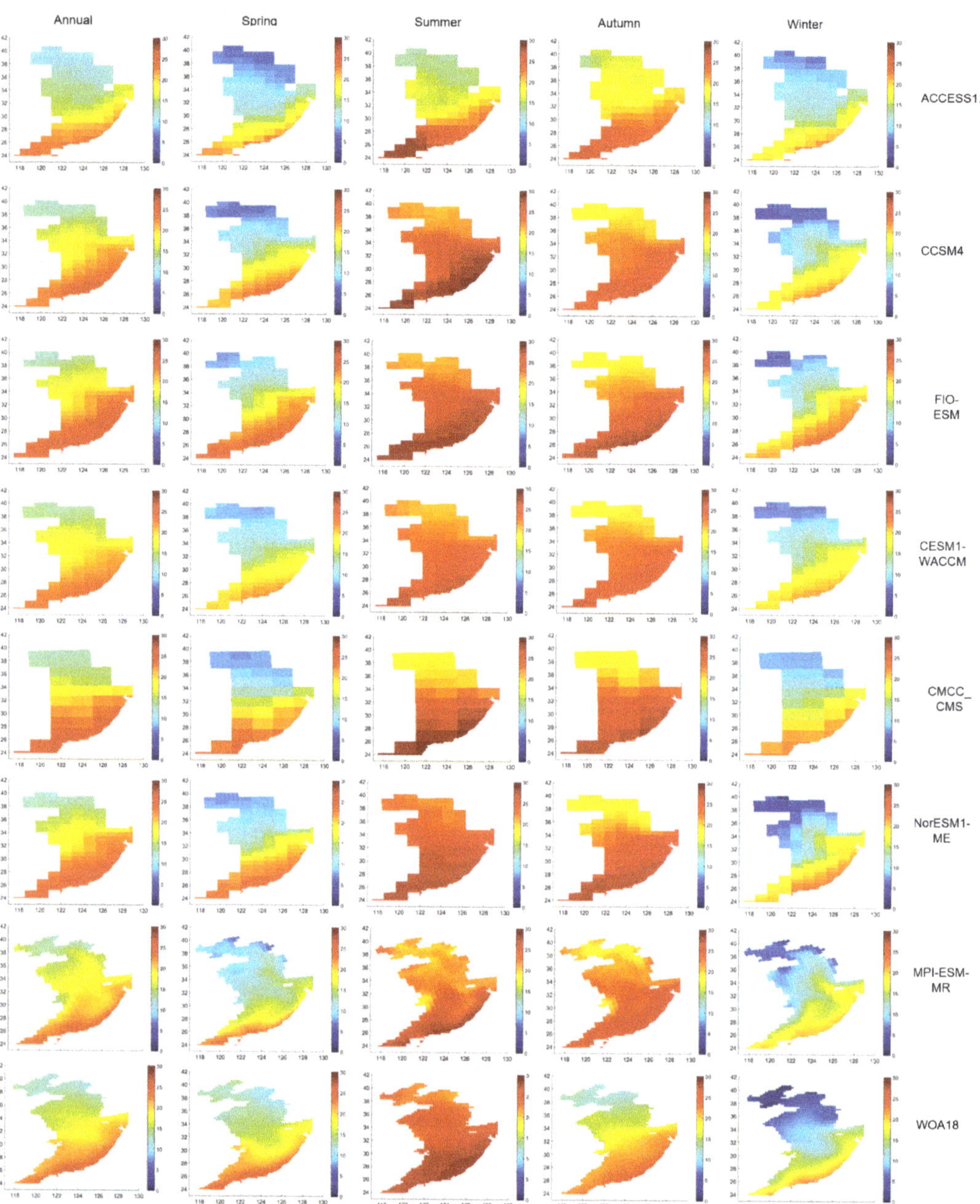

Figure 2. Model validation: the comparison of CMIP5 simulation results with WOA18. The first row to seventh row illustrate the result of seven CMIP5 models: ACCESS1.3, CCSM4, FIO-ESM, CESM1-CAM5, CMCC-CMS, NorESM1-ME, and MPI-ESM-MR, respectively. The last row shows the observation result from WOA18 data. The first to fifth column present the results in the year round, spring, summer, autumn, and winter of 2010, respectively.

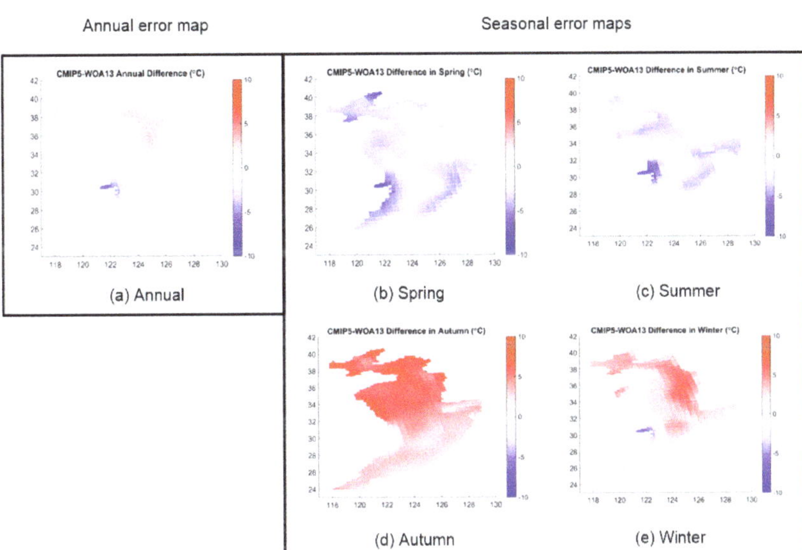

Figure 3. The error maps of MPI-ESM-MR in 2010: (**a**) annual, (**b**) spring, (**c**) summer, (**d**) autumn, and (**e**) winter.

4. Projection Results of Decadal SST Changes

We analyzed the decadal SST changes in 2030, 2060, and 2090 compared with the present analysis in 2010. The modified projection results of the seven models and their average results are shown in Table 3 and Figure 4, respectively. From Table 3 and Figure 4, we find that most models demonstrate a significant SST increase in the next 100 years. From Table 3, we also find that the simulation results of MPI-ESM-MR are most similar to the average of all models. Moreover, its resolution is much higher than that of other models. Since the resolution of the other six models is too low to distinguish local climate impact factors such as the Kuroshio system, it is inappropriate to use them to project the SST changes in the future. Therefore, next, we used the modified result of MPI-ESM-MR to analyze the decadal SST changes. From Table 3, we can see that the SST increases from 18.09 °C in 2010 to 18.68 °C in 2030, 19.78 °C in 2060, and 19.65 °C in 2090. In particular, a remarkable SST increase was obtained from 2030 to 2060, which was 1.10 °C. Till 2090, the SST will increase by 1.55 °C. Besides the decadal analysis, we also investigated the annual changing rate of SST from 2010 to 2090. The changing rate between 2030 and 2010, 2060 and 2030, and 2090 and 2060 is illustrated in Table 3. We can see that this rate is very small from 2010 to 2030, achieves the top value from 2030 to 2060, and becomes low from 2060 to 2090. This suggests that the SST increase becomes stable from 2060 to 2090.

Then, we analyzed the spatial SST changes in different regions through the qualitative results shown in Figure 5. Figure 5a,c,e shows the SST projection in 2030, 2060, and 2090 and their changes compared to 2010 in (b), (d), and (f), respectively. From Figure 5, we can find that the highest SST increase rate is obtained in 2060 (in Figure 5d), and the increase becomes stable in 2090 (in Figure 5f). Although the SST increase in 2030 (in Figure 5b) is not as significant as that in 2060, its distribution is inhomogeneous. The significant SST increases mainly concentrate in the east of Bohai (especially east of Qinhuangdao), the Changjiang Estuary, and its adjacent shore of Jiangsu Province. The increment can reach to as high as 1.5 °C. From Figure 5d, we can see that compared to 2030, the increase of the SST in 2060 expands to almost all the regions and at an even higher rate. The increment reaches about 3.5 °C in the Changjiang Estuary and in the outer shelf of the East China Sea. From Figure 5f, we can see that the SST variation in 2090 is similar to that of 2060 and becomes stable gradually. One remarkable thing to note is that the SST in the northern and central Yellow Sea is relatively lower than that in the other regions (see Figure 5a,c,e), and their SST increases are not very obvious (see Figure 5b). This is mainly due to the presence of the cold water mass in the Northern and Central Yellow Seas, which has a great influence on the distribution of the SST. The increase of the SST can severely influence marine ecosystems and cause many ecological problems, especially for shelf regions with shallow water. For example, the Brown tide has broken out in the Qinhuangdao coastal area of Bohai recurrently since 2009. The Changjiang Estuary and the shore in Northern Jiangsu Province suffered from the Red Tide and Green Tide, and jellyfish disasters often have also occurred in these areas recently. The increase of the SST will intensify these problems through the following mechanisms. First, it can affect the metabolic rate of marine organisms. Second, it can influence the other oceanic states, such as local currents. Third, it can further affect the water column stratification, substrate structure, photosynthetic light intensity, and nutrient cycling [3]. The species distribution can also be disturbed by the increase of the SST. For example, the warm and cold water fish stocks can be reduced greatly in the regions with significant SST increases, and tropical ocean organisms may move up to the middle and high latitude areas with warm water [37].

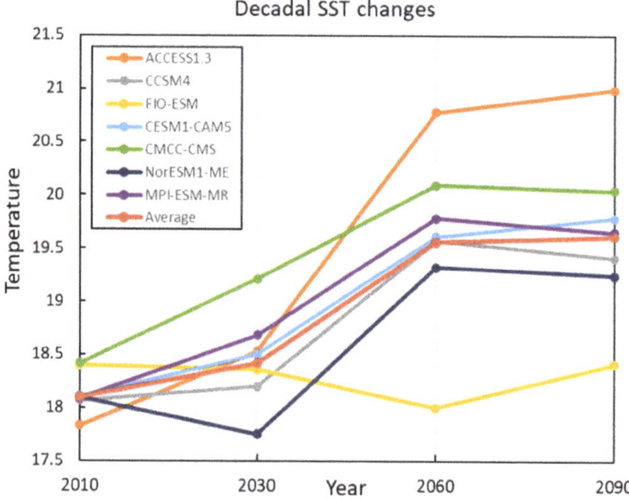

Figure 4. The decadal SST variations in the next 100 years.

Table 3. The decadal SST variations in the next 100 years.

SST	2010	2030	2060	2090
ACCESS1.3	17.83	18.53	20.78	20.99
CCSM4	18.07	18.20	19.57	19.41
FIO-ESM	18.40	18.36	18.00	18.41
CESM1-CAM5	18.11	18.50	19.61	19.78
CMCC-CMS	18.42	19.21	20.09	20.04
NorESM1-ME	18.10	17.75	19.32	19.24
MPI-ESM-MR	18.09	18.68	19.78	19.65
Average	18.10	18.42	19.56	19.61

Table 3. *Cont.*

SST Changes	2010	2030–2010	2060–2010	2090–2010
		(2030–2010)/Years	(2060–2030)/Years	(2090–2060)/Years
ACCESS1.3	17.83	0.70	2.95	3.16
		0.033	0.073	0.007
CCSM4	18.07	0.13	1.50	1.34
		0.006	0.044	−0.005
FIO-ESM	18.40	−0.04	−0.40	0.02
		−0.002	−0.011	0.013
CESM1-CAM5	18.11	0.40	1.50	1.67
		0.019	0.036	0.005
CESM1-CMS	18.42	0.79	1.67	1.62
		0.038	0.028	−0.002
NorESM1-ME	18.10	−0.35	1.22	1.14
		−0.016	0.051	−0.003
MPI-ESM-MR	18.10	0.59	1.68	1.55
		0.028	0.035	−0.004
Average	18.11	0.32	1.46	1.51
		0.014	0.036	0.002

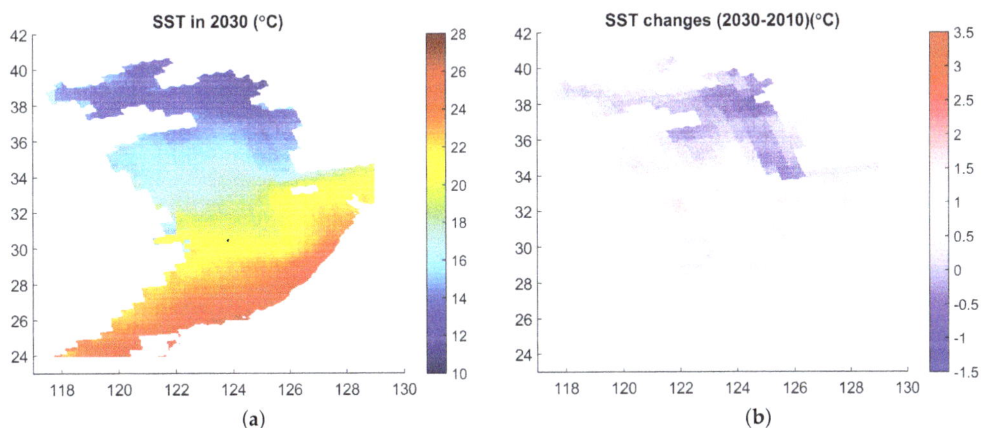

Figure 5. *Cont.*

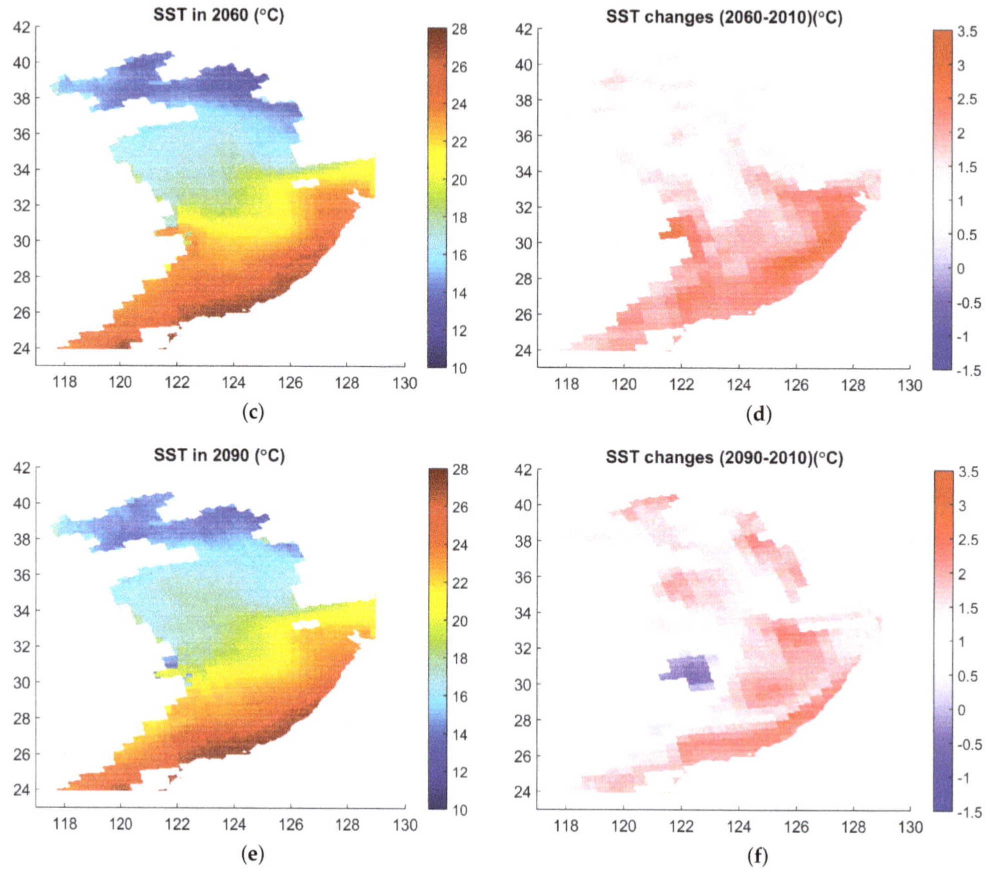

Figure 5. The projection results of decadal SST changes by MPI-ESM-MR in (**a**) 2030, (**c**) 2060, and (**e**) 2090 and their changes compared to that in 2010 in (**b**) 2030–2010, (**d**) 2060–2010, and (**f**) 2090–2010.

5. Projection Results of Seasonal SST Changes

We analyzed the seasonal SST variations in the next 100 years. The numerical results of the seven models and their average results are shown in Table 4. Similar to the decadal analysis, we used the modified simulation result of MPI-ESM-MR to investigate the seasonal SST changes because of its high resolution and low projection error. From Table 4, we can see that the SST increases significantly from 2010 to 2090 for all seasons. The increment reaches about 2 °C by 2090. Next, we analyzed the spatial seasonal SST variations. The qualitative simulation results are shown in Figure 6. The first to fourth rows show the SST changes from 2010 to 2030, 2060, and 2090 in spring, summer, autumn, and winter, respectively. We found that in summer, the regions with significant SST increases mainly are concentrated in the east of Qinhuangdao in Bohai, the shore of Northern Jiangsu Province in the Yellow Sea, and the Changjiang Estuary in the East China Sea, which is coincident with that in the decadal analysis. This is mainly because the upwelling waters become weak and the Kuroshio invasion becomes stronger in summer. In autumn, the increase of the SST becomes smoother, and the regions with a significant SST increase move from the inner shelf area to the middle and even outer shelf regions in winter. In particular, the outer shelf regions have higher SST increases than other regions. A remarkable demonstration of the SST increase in the Northern Yellow Sea is illustrated in Figure 6f, where the Yellow

Sea warm current plays a more significant role in summer than the Yellow Sea cold water mass. The SST increases in this region will intensify the low oxygen and acidification problems. Similar to the decadal analysis, for all seasons, the SST increase in 2060 (shown in the second column) is more significant than that in 2030 (shown in the first column) and becomes stable in 2090 (shown in the third column). As previously mentioned, the increase of the SST may cause many environmental problems and destroy the marine ecosystem. For example, the increase of the SST in Bohai may lead to the Brown Tide problem, and the SST increases in the shore of Jiangsu Province and the Changjiang Estuary may result in the Red and Green Tide problem and even cause the outbreak of jellyfish disasters in these regions. The increase of the SST in the middle shelf and outer shelf regions of the East China Sea will intensify the marine acidification and ocean hypoxia there. The model validation and modification of this paper is relative simple. We use the difference between the CMIP5 model and the WOA data in 2010 as the model error and substrate these errors from CMIP5 model data to modify their projection results. We should use more sophisticated methods with all the model assumptions and conditions to perform model modification. However, it is not an easy task especially for the future projection with many unknown conditions. We donot know whether these errors will remain constant for the 2010–2100 period. Thus. we include the experimental results without projection modification for comparison in this part. The decadal and seasonal SST changes without modification of model projection results are shown in Table A1 and Table A2, respectively. Comparing Table A1 to Table 3, and Table A2 to Table 4, we find that the bottom part of these tables are the same with each other. That is, the investigation of 35 the SST changes in the future using the differences of 2030, 2060, and 2090 relative to that in 2010 is not influenced by the 36 model validation and modification method in this paper.

Table 4. The projection results of seasonal SST variations in the next 100 years using seven CMIP5 models and their average result.

SST	2010				2030				2060				2090			
	Spr.	Sum.	Aut.	Win.	Spr.	Sum.	Aut.	Win.	Spr.	Sum.	Aut.	Win.	Spr.	Sum.	Aut.	Win.
ACCESS1.3	16.20	24.86	18.77	11.48	16.55	25.38	19.63	12.57	19.47	28.28	21.75	13.64	18.78	28.49	22.65	14.05
CCSM4	16.43	25.71	18.85	11.29	16.45	25.36	19.57	11.44	18.04	26.61	20.53	13.08	17.83	26.64	20.68	12.49
FIO-ESM	16.94	25.84	19.06	11.75	16.36	26.20	19.26	11.60	15.52	25.95	19.78	10.77	16.45	26.68	19.07	11.47
CESM1-CAM5	16.48	25.64	18.88	11.42	17.33	26.21	19.03	11.45	18.65	27.36	19.98	12.46	17.94	27.54	21.02	12.61
CMCC-CMS	16.76	25.60	19.32	11.99	18.10	26.42	19.58	12.75	18.83	27.64	20.46	13.44	18.83	27.39	20.10	13.85
NorESM1-ME	16.54	25.77	18.85	11.24	16.80	25.73	18.18	10.30	17.93	27.68	19.72	11.96	17.96	27.36	19.52	12.13
MPI-ESM-MR	16.66	25.71	18.73	11.26	17.36	26.88	18.41	12.06	19.17	27.12	19.71	13.10	18.27	27.76	19.76	12.88
Average	16.67	25.71	18.72	11.31	17.08	26.15	18.88	11.58	18.34	27.35	20.07	12.48	18.11	27.53	20.19	12.61

SST Changes	2010				2030–2010				2060–2010				2090–2010			
	Spr.	Sum.	Aut.	Win.	Spr.	Sum.	Aut.	Win.	Spr.	Sum.	Aut.	Win.	Spr.	Sum.	Aut.	Win.
ACCESS1.3	16.20	24.86	18.77	11.48	0.35	0.52	0.86	1.09	3.27	3.41	2.97	2.16	2.58	3.63	3.87	2.57
CCSM4	16.43	25.71	18.85	11.29	0.02	−0.36	0.72	0.15	1.61	0.90	1.68	1.79	1.40	0.93	1.82	1.20
FIO-ESM	16.94	25.84	19.06	11.75	−0.57	0.36	0.20	−0.15	−1.42	0.11	0.72	−0.99	−0.49	0.84	0.00	−0.28
CESM1-CAM5	16.48	25.64	18.88	11.42	0.85	0.57	0.15	0.03	2.16	1.72	1.10	1.04	1.45	1.90	2.14	1.19
CMCC-CMS	16.76	25.60	19.32	11.99	1.33	0.81	0.26	0.75	2.07	2.04	1.15	1.45	2.06	1.79	0.79	1.86
NorESM1-ME	16.54	25.77	18.85	11.24	0.27	−0.05	−0.68	−0.93	1.39	1.91	0.87	0.73	1.42	1.59	0.67	0.89
MPI-ESM-MR	16.66	25.71	18.73	11.26	0.70	1.17	−0.33	0.80	2.51	1.41	0.98	1.84	1.61	1.95	1.03	1.62
Average	16.67	25.71	18.72	11.31	0.41	0.44	0.16	0.26	1.67	1.64	1.35	1.16	1.45	1.82	1.47	1.29

SST Changes ratios	2010				2030–2010				2060–2030				2090–2060			
	Spr.	Sum.	Aut.	Win.	Spr.	Sum.	Aut.	Win.	Spr.	Sum.	Aut.	Win.	Spr.	Sum.	Aut.	Win.
ACCESS1.3	16.20	24.86	18.77	11.48	0.017	0.025	0.041	0.052	0.094	0.093	0.068	0.035	0.022	0.007	0.029	0.013
CCSM4	16.43	25.71	18.85	11.29	0.001	−0.017	0.034	0.007	0.051	0.041	0.031	0.031	0.053	0.007	0.008	0.005
FIO-ESM	16.94	25.84	19.06	11.75	−0.027	0.017	0.009	−0.007	−0.027	−0.008	0.017	−0.027	−0.030	0.024	−0.023	0.023
CESM1-CAM5	16.48	25.64	18.88	11.42	0.040	0.027	0.007	0.001	0.042	0.037	0.031	0.033	0.023	0.006	0.034	0.005
CMCC-CMS	16.76	25.60	19.32	11.99	1.33	0.81	0.26	0.75	2.07	2.04	1.15	1.45	2.06	1.79	0.79	1.86
NorESM1-ME	16.54	25.77	18.85	11.24	0.013	−0.002	0.032	0.045	0.036	0.063	0.050	0.054	0.008	−0.010	−0.007	0.005
MPI-ESM-MR	16.66	25.71	18.73	11.26	0.034	0.056	−0.016	0.038	0.058	0.008	0.042	0.034	−0.029	0.017	0.002	−0.007
Average	16.67	25.71	18.72	11.31	0.020	0.021	0.007	0.013	0.041	0.039	0.038	0.029	−0.007	0.006	0.004	0.004

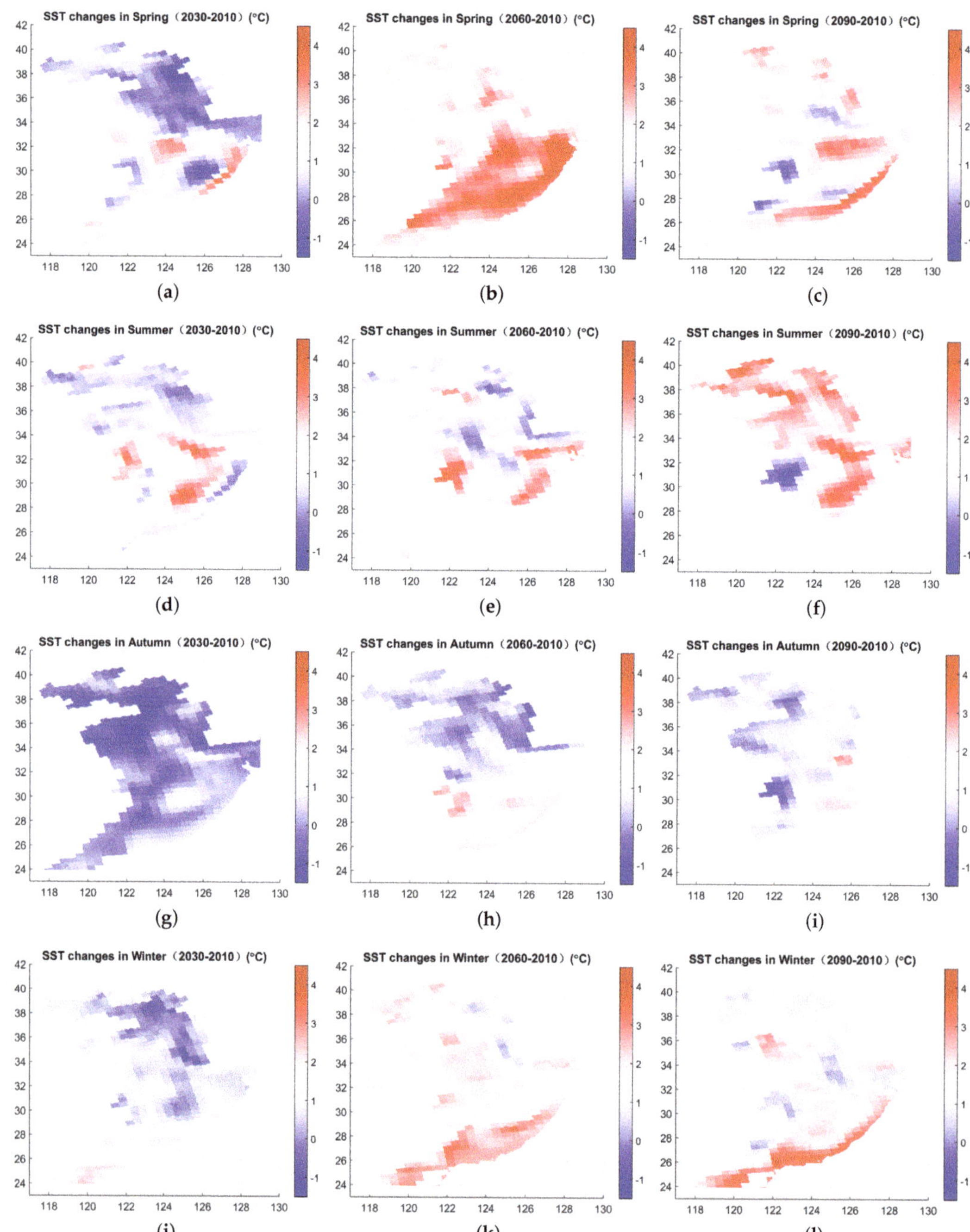

Figure 6. Seasonal changes of SST from 2010 to 2030 (the first column (**a,d,g,j**)), 2060 (the second column (**b,e,h,k**)), and 2090 (the third column (**c,f,i,l**)), respectively. The first row to fourth row illustrates the changes in Spring, Summer, Autumn, and Winter, respectively.

6. Conclusions

This paper investigated both decadal and seasonal SST variations in the East China Shelf Seas by 2100 using CMIP5 models under RCP4.5. Seven representative CMIP5 models were used in this paper, including ACCESS1.3, CCSM4, FIO-ESM, CESM1-CAM5, CMCC-CMS, NorESM1-ME, and MPI-ESM-MR. To make the projection more accurate, we introduced the high-resolution hydrological observation data from WOA18 for model validation and simulation result modification. The best model (MPI-ESM-MR) with highest resolution and low errors was selected for the projection of SST changes. The extensive experimental results demonstrated that both decadal and seasonal SSTs in the East China Shelf Seas will increase significantly in the next one hundred years: the decadal SST will increase by about 1.5 °C till 2090, and the seasonal SST will increase by about 1.03 °C–1.95 °C by 2090. The highest increment was obtained from 2030 to 2060, and it became stable from 2060 to 2090. Although the SST increase in 2030 was not as significant as that in 2060 and 2090, the distribution of SST increases was inhomogeneous. Some local regions had a high SST increase of 1.5 °C in the east of Bohai Sea (east of Qinhuangdao), the Changjiang Estuary, and its adjacent shore of Jiangsu Province. Compared to the decadal analysis, the seasonal variation may play a more important role in understanding climate changes. We found that in summer, the significant SST increase mainly concentrated in the eastern area of Bohai, the Changjiang Estuary, and the shore of Jiangsu Province, which is mainly due to the weakness of upwelling, the strengthening of the Kuroshio invasion, and the influence of the Taiwan warm current. In autumn, the increase of SST becomes smoother, and the regions with a significant SST increase move from the inner shelf to the outer shelf regions in winter and spring. The significant SST increase may affect the local marine ecology seriously. It can change the distribution of marine organisms, lead to the moving up of the phenophase, intensify the low oxygen and acidification problems, and cause many environmental problems. It has brought about the Brown Tide in the east of Qinhuangdao in the Bohai Sea, led to the Red and Green Tide in the Changjiang Estuary and the shore of Jiangshu Province, and caused the outbreak of jellyfish disasters in these regions. To the best of our knowledge, this is the first work to investigate both decadal and seasonal SST variations in the East China Shelf Seas using high-resolution CMIP5 model data and observation data.

Author Contributions: Conceptualization, C.Z.; Methodology, C.Z., C.X., H.L. and J.Z.; Analysis, C.Z., C.X. and H.L.; Writing and Editing, C.Z., H.L.; Project Administration, C.Z. and J.Z. All authors have read and agreed to the published version of the manuscript.

Funding: This research was funded by the National Natural Science Foundation of China Grant 41806116 and the National Key Research and Development Program of China 2016YFC1401203.

Institutional Review Board Statement: Not applicable.

Informed Consent Statement: Not applicable.

Data Availability Statement: Data used can be accesses via CMIP5 website.

Acknowledgments: H.L. acknowledges the National Natural Science Foundation of China Grant 41806116 and the National Key Research and Development Program of China 2016YFC1401203.

Conflicts of Interest: The authors declare no conflict of interest.

Appendix A

Table A1. The decadal SST variation in the next 100 years using the raw CMIP5 model data without projection result modification.

SST	2010	2030	2060	2090
ACCESS1.3	16.18	16.89	19.14	19.35
CCSM4	18.93	19.06	20.42	20.26
FIO-ESM	20.48	20.44	20.08	20.50
CESM1-CAM5	18.84	19.24	20.34	20.51
CMCC-CMS	19.47	20.26	21.14	21.10
NorESM1-ME	19.47	19.12	20.69	20.61
MPI-ESM-MR	18.51	19.10	20.19	20.06
Average	18.81	19.13	20.27	20.32

SST Changes	2010	2030–2010	2060–2010	2090–2010
		(2030–2010)/Years	(2060–2030)/Years	(2090–2060)/Years
ACCESS1.3	17.83	0.70	2.95	3.16
		0.034	0.073	0.007
CCSM4	18.07	0.13	1.50	1.34
		0.006	0.044	−0.005
FIO-ESM	18.40	−0.04	−0.40	0.02
		−0.002	−0.011	0.013
CESM1-CAM5	18.11	0.40	1.50	1.67
		0.019	0.036	0.005
CESM1-CMS	18.42	0.79	1.67	1.62
		0.038	0.028	−0.002
NorESM1-ME	18.10	−0.35	1.22	1.14
		−0.017	0.051	−0.003
MPI-ESM-MR	18.10	0.59	1.68	1.55
		0.028	0.035	−0.004
Average	18.11	0.32	1.46	1.51
		0.015	0.037	0.002

The model validation and modification of this paper is relative simple. We use the difference between the CMIP5 model and the WOA data in 2010 as the model error and substrate these errors from CMIP5 model data to modify their projection results. We should use more sophisticated methods with all the model assumptions and conditions to perform model modification. However, it is not an easy task especially for the future projection with many unknown conditions. We donot know whether these errors will remain constant for the 2010–2100 period. Thus. we include the experimental results without projection modification for comparison in this part. The decadal and seasonal SST changes without modification of model projection results are shown in Table A1 and Table A2, respectively. Comparing Table A1 to Table 3, and Table A2 to Table 4, we find that the bottom part of these tables are the same with each other. That is, the investigation of 35 the SST changes in the future using the differences of 2030, 2060, and 2090 relative to that in 2010 is not influenced by the 36 model validation and modification method in this paper.

Table A2. The projection results of seasonal SST variations in the next 100 years using seven CMIP5 models and their average result using the raw CMIP5 model data without projection result modification.

SST	2010				2030				2060				2090			
	Spr.	Sum.	Aut.	Win.	Spr.	Sum.	Aut.	Win.	Spr.	Sum.	Aut.	Win.	Spr.	Sum.	Aut.	Win.
ACCESS1.3	12.55	24.86	20.28	13.00	12.89	19.43	21.14	14.09	15.81	22.33	23.25	15.16	15.13	22.54	24.15	15.57
CCSM4	14.41	25.34	22.79	13.16	14.43	24.98	23.51	13.31	16.02	26.24	24.47	14.95	15.81	26.27	24.62	14.36
FIO-ESM	17.40	25.51	23.48	15.54	16.82	25.87	23.67	15.39	15.98	25.62	24.20	14.56	16.91	26.35	23.47	15.26
CESM1-CAM5	14.27	23.98	22.92	14.17	15.12	24.55	23.07	14.20	16.43	25.70	24.03	15.21	15.72	25.88	25.07	15.36
CMCC-CMS	15.46	23.88	23.52	15.03	16.78	24.69	23.78	15.78	17.52	25.92	24.66	16.47	17.52	25.67	24.30	16.89
NorESM1-ME	15.49	25.63	23.55	13.20	15.76	25.59	22.87	12.27	16.89	27.54	24.41	13.93	16.91	27.36	19.52	12.13
MPI-ESM-MR	14.06	23.53	23.25	13.20	14.76	24.70	22.92	14.00	16.57	24.94	24.23	15.04	15.67	25.48	24.27	14.82
Average	14.76	23.81	22.83	13.85	15.17	24.25	22.99	14.11	16.43	25.45	24.18	15.01	16.20	25.63	24.30	15.14

SST Changes	2010				2030–2010				2060–2010				2090–2010			
	Spr.	Sum.	Aut.	Win.	Spr.	Sum.	Aut.	Win.	Spr.	Sum.	Aut.	Win.	Spr.	Sum.	Aut.	Win.
ACCESS1.3	12.55	24.86	20.28	13.00	0.35	0.52	0.86	1.09	3.27	3.41	2.97	2.16	2.58	3.63	3.87	2.57
CCSM4	14.41	25.34	22.79	13.16	0.02	−0.36	0.72	0.15	1.61	0.90	1.68	1.79	1.40	0.93	1.82	1.20
FIO-ESM	17.40	25.51	23.48	15.54	−0.57	0.36	0.20	−0.15	−1.42	0.11	0.72	−0.99	−0.49	0.84	0.00	−0.28
CESM1-CAM5	14.27	23.98	22.92	14.17	0.85	0.57	0.15	0.03	2.16	1.72	1.10	1.04	1.45	1.90	2.14	1.19
CMCC-CMS	15.46	23.88	23.52	15.03	1.33	0.81	0.26	0.75	2.07	2.04	1.15	1.45	2.06	1.79	0.79	1.86
NorESM1-ME	15.49	25.63	23.55	13.20	0.27	−0.05	−0.68	−0.93	1.39	1.91	0.87	0.73	1.42	1.59	0.67	0.89
MPI-ESM-MR	14.06	23.53	23.25	13.20	0.70	1.17	−0.33	0.80	2.51	1.41	0.98	1.84	1.61	1.95	1.03	1.62
Average	14.76	23.81	22.83	13.85	0.41	0.44	0.16	0.26	1.67	1.64	1.35	1.16	1.45	1.82	1.47	1.29

SST Changes ratios	2010				2030–2010				2060–2030				2090–2060			
	Spr.	Sum.	Aut.	Win.	Spr.	Sum.	Aut.	Win.	Spr.	Sum.	Aut.	Win.	Spr.	Sum.	Aut.	Win.
ACCESS1.3	12.55	24.86	20.28	13.00	0.017	0.025	0.041	0.052	0.094	0.093	0.068	0.035	−0.022	0.007	0.029	0.013
CCSM4	14.41	25.34	22.79	13.16	0.001	−0.017	0.034	0.007	0.051	0.041	0.031	0.053	−0.007	0.001	0.005	−0.019
FIO-ESM	17.40	25.51	23.48	15.54	−0.027	0.017	0.009	−0.007	−0.027	−0.008	0.017	−0.027	−0.030	0.024	−0.023	0.023
CESM1-CAM5	14.27	23.98	22.92	14.17	0.040	0.027	0.007	0.001	0.042	0.037	0.031	0.033	−0.023	0.006	0.034	0.005
CMCC-CMS	15.46	23.88	23.52	15.03	0.063	0.039	0.013	0.036	0.024	0.040	0.029	0.022	−0.001	−0.008	−0.012	0.013
NorESM1-ME	15.49	25.63	23.55	13.20	0.013	−0.002	0.032	0.045	0.036	0.063	0.050	0.054	0.001	−0.010	−0.007	0.005
MPI-ESM-MR	14.06	23.53	23.25	13.20	0.034	0.056	−0.016	0.038	0.058	0.008	0.042	0.034	−0.029	0.017	0.002	−0.007
Average	14.76	23.81	22.83	13.85	0.020	0.021	0.007	0.013	0.041	0.039	0.038	0.029	−0.007	0.006	0.004	0.004

References

1. Cheng, L.; Abraham, J.; Hausfather, Z.; Trenberth, K.E. How fast are the oceans warming? *Science* **2019**, *363*, 128–129. [CrossRef] [PubMed]
2. Wu, R.; Lin, J.; Li, B. Spatial and temporal variability of sea surface temperature in Eastern Marginal Seas of China. *Adv. Meteorol.* **2016**, *2*, 128–129. [CrossRef]
3. Lin, C.; Su, J.; Xu, B.; Tang, Q. Long-term variations of temperature and salinity of the Bohai sea and their influence on its ecosystem. *Prog. In Oceanogr.* **2001**, *49*, 7–19. [CrossRef]
4. Chu, P.; Chen, Y.; Kuninaka, A. Seasonal variability of the Yellow Sea/East China Sea surface fluxes and thermohaline structure. *Adv. Atmos. Sci.* **2005**, *49*, 1–20. [CrossRef]
5. Liu, K.K.; Chao, S.Y.; Lee, H.J.; Gong, G.C.; Teng, Y.C. Seasonal variation of primary productivity in the East China Sea: A numerical study based on coupled physical-biogeochemical model. *Deep-Sea Res. Part II* **2010**, *57*, 1762–1782. [CrossRef]
6. Yu, X.; Wang, F.; Tang, X. Future projection of East China Sea temperature by dynamic downscaling of the IPCC-AR4 CCSM3 model result. *Chin. J. Oceanol. Limnol.* **2012**, *30*, 826–842. [CrossRef]
7. Yuan, Z.; Xiao, X.; Wang, F.; Xing, L.; Wang, Z.; Zhang, H.; Xiang, R.; Zhou, L.; Zhao, M. Spatiotemporal temperature variations in the East China Sea shelf during the Holocene in response to surface circulation evolution. *Quat. Int.* **2018**, *482*, 46–55. [CrossRef]
8. Tang, X.; Wang, F.; Chen, Y.; Mingkui, L.I. Warming trend in northern East China Sea in recent four decades. *CJOL* **2009**, *27*, 185–191. [CrossRef]
9. Wu, R.; Li, C.; Lin, J. Enhanced winter warming in the Eastern China Coastal Waters and its relationship with ENSO. *Atmos. Sci. Lett.* **2017**, *18*, 11–18. [CrossRef]
10. Solomon, S. IPCC (2007): Climate change the physical science basis. In Proceedings of the AGU Fall Meeting, San Francisco, CA, USA, 10–14 December 2007; pp. 123–124.

11. Stocker, T.F.; Qin, D.; Plattner, G.K.; Tignor, M.M.B.; Allen, S.K.; Boschung, J.; Nauels, A.; Xia, Y.; Bex, V.; Midgley, P.M. Climate change 2013: The physical science basis. In Proceedings of the AGU Fall Meeting, San Francisco, CA, USA, 15–19 December 2014; pp. 1022–1025.

12. Cheng, L.; Zhu, J. Benefits of CMIP5 multimodel ensemble in reconstructing historical ocean subsurface temperature variations. *J. Clim.* **2016**, *29*, 5393–5416. [CrossRef]

13. Zhou, B.T.; Ying, X.U. CMIP5 analysis of the interannual variability of the pacific SST and its association with the Asian-Pacific oscillation. *Atmos. Ocean. Sci. Lett.* **2017**, *10*, 138–145. [CrossRef]

14. Qu, X.; Huang, G. The decadal variability of the tropical Indian Ocean SST–the South Asian High relation: CMIP5 model study. *Clim. Dyn.* **2015**, *45*, 273–289. [CrossRef]

15. Qin, P.; Xie, Z. Precipitation extremes in the dry and wet regions of China and their connections with the sea surface temperature in the eastern tropical Pacific Ocean. *J. Geophys. Res. D Atmos. JGR* **2017**, *122*, 6273–6283. [CrossRef]

16. Tachibana, Y.; Iizuka, S.; Kodama, Y.-M.; Moteki, Q.; Manda, A.; Yamada, H.; Ishida, S.; Fathrio, I. Assessment of western Indian Ocean SST bias of CMIP5 models. *J. Geophys. Res. C Ocean. JGR* **2017**, *122*, 3123–3140.

17. Song, Z.Y.; Liu, H.L.; Wang, C.Z.; Zhang, L.P.; Qiao, F.L. Evaluation of the eastern equatorial Pacific SST seasonal cycle in CMIP5 models. *Ocean. Sci.* **2014**, *10*, 837–843. [CrossRef]

18. Xu, Z.; Chang, P.; Richter, I.; Tang, G. Diagnosing southeast tropical Atlantic SST and ocean circulation biases in the CMIP5 ensemble. *Clim. Dyn. Obs. Theor. Comput. Res. Clim. Syst.* **2014**, *43*, 3123–3145. [CrossRef]

19. Zhao, Y.; Zhang, H. Impacts of SST warming in tropical Indian Ocean on CMIP5 model-projected summer rainfall changes over Central Asia. *Clim. Dyn. Obs. Theor. Comput. Clim. Syst.* **2016**, *46*, 3223–3238. [CrossRef]

20. Kucharski, F.; Joshi, M.K. Influence of tropical South Atlantic sea surface temperatures on the Indian summer monsoon in CMIP5 models: Tropical South Atlantic influence on the Indian summer monsoon, Quarterly. *J. R. Meteorol. Soc.* **2017**, *143*, 5393–5416. [CrossRef]

21. Levine, R.C.; Turner, A.G.; Marathayil, D.; Martin, G.M. The role of northern Arabian Sea surface temperature biases in CMIP5 model simulations and future projections of Indian summer monsoon rainfall. *Clim. Dyn.* **2013**, *41*, 155–172. [CrossRef]

22. Langehaug, H.R.; Matei, D.; Eldevik, T.; Lohmann, K.; Gao, Y. Decadal predictability of SST in the Atlantic domain of the Nordic Seas in three CMIP5 models. In Proceedings of the EGU General Assembly Conference, Vienna, Austria, 12–17 April 2015.

23. Huang, C.; Qiao, F.; Song, Y.; Li, X. The simulation and forecast of SST in the South China Sea by CMIP5 models. *Acta Oceanol. Sin.* **2014**, *36*, 38–47. (In Chinese)

24. Tan, H.; Cai, R.; Yan, X. Projected 21st century sea surface temperature over offshore china based on IPCC-CMIP5 models. *J. Appl. Oceanogr.* **2016**, *35*, 451–458.

25. Song, C.; Zhang, S.; Jiang, H.; Wang, H.; Wang, D.; Huang, Y. Evaluation and projection of SST in the China seas from CMIP5. *Acta Oceanol. Sin.* **2016**, *38*, 1–11. (In Chinese)

26. Locarnini, M.; Mishonov, A.V.; Baranova, O.K. *World Ocean Atlas 2018, Volume 1: Temperature*; Levitus, S., Mishonov, A., Eds.; NOAA Atlas NESDIS 81; NOAA/NESDIS National Centers for Environmental Information: Silver Spring, MD, USA, 2018; 52p.

27. Heuzé, C.; Heywood, K.J.; Stevens, D.P.; Ridley, J.K. Southern Ocean bottom water characteristics in CMIP5 models. *Geophys. Res. Lett.* **2013**, *40*, 1409–1414. [CrossRef]

28. Semenov, V.A.; Martin, T.; Behrens, L.K.; Latif, M. Arctic sea ice area in CMIP3 and CMIP5 climate model ensembles—Variability and change. *Cryosphere Discuss.* **2015**, *9*, 1077–1131.

29. Giorgetta, M.A.; Jungclaus, J.; Reick, C.H.; Legutke, S.; Bader, J.; Böttinger, M.; Brovkin, V.; Crueger, T.; Esch, M.; Fieg, K.; et al. Climate and carbon cycle changes from 1850 to 2100 in MPI-ESM simulations for the Coupled Model Intercomparison Project phase 5. *J. Adv. Model. Earth Syst.* **2013**, *5*, 572–597. [CrossRef]

30. Bi, D.; Dix, M.; Marsland, S.; O'Farrell, S.; Rashid, H.; Uotila, P.; Hirst, A.; Kowalczyk, E.; Golebiewski, M.; Sullivan, A.; et al. The ACCESS coupled model: Description, control climate and evaluation. *Aust. Meteorol. Oceanogr. J.* **2013**, *63*, 41–64. [CrossRef]

31. Danabasoglu, G.; Bates, S.C.; Briegleb, B.P.; Jayne, S.R.; Jochum, M.; Large, W.G.; Peacock, S.; Yeager, S.G. The CCSM4 ocean component. *J. Clim.* **2012**, *25*, 1361–1389. [CrossRef]

32. Qiao, F.; Song, Z.; Bao, Y.; Song, Y.; Shu, Q.; Huang, C.; Zhao, W. Development and evaluation of an Earth System Model with surface gravity waves. *J. Geophys. Res. Ocean.* **2013**, *118*, 4514–4524. [CrossRef]

33. Meehl, G.A.; Washington, W.M.; Arblaster, J.M.; Hu, A.; Teng, H.; Kay, J.E.; Gettelman, A.; Lawrence, D.M.; Sanderson, B.M.; Strand, W.G. Climate change projections in CESM1(CAM5) compared to CCSM4. *J. Clim.* **2013**, *26*, 6287–6308. [CrossRef]

34. Borrelli, A.; Materia, S.; Bellucci, A.; Gualdi, S. Seasonal prediction system at CMCC. *SSRN Electron. J.* **2012**, *147*, 1–18. [CrossRef]

35. Schwinger, J.; Goris, N.; Tjiputra, J.F.; Kriest, I.; Bentsen, M.; Bethke, I.; Ilicak, M.; Assmann, K.M.; Heinze, C. Evaluation of NorESM-OC (versions 1 and 1.2), the ocean carbon-cycle stand-alone configuration of the Norwegian Earth System Model (NorESM1). *Geosci. Model Dev.* **2016**, *9*, 1–73. [CrossRef]

36. Jungclaus, J.H.; Fischer, N.; Haak, H.; Lohmann, K.; Marotzke, J.; Matei, D.; Mikolajewicz, U.; Notz, D.; Storch, J.S.V. Characteristics of the ocean simulations in the Max Planck Institute Ocean Model (MPIOM) the ocean component of the MPI-Earth system model. *J. Clim.* **2013**, *5*, 422–446. [CrossRef]

37. Yan, X.; Cai, R.; Bai, Y. Long-term change of the marine environment and plankton in the Xiamen Sea under the influence of climate change and human sewage. *Toxicol. Environ. Chem. Rev.* **2015**, *98*, 669–678. [CrossRef]

MDPI

St. Alban-Anlage 66

4052 Basel

Switzerland

Tel. +41 61 683 77 34

Fax +41 61 302 89 18

www.mdpi.com

Journal of Marine Science and Engineering Editorial Office

E-mail: jmse@mdpi.com

www.mdpi.com/journal/jmse

MDPI

St. Alban-Anlage 66

4052 Basel

Switzerland

Tel. +41 61 683 77 34

Fax +41 61 302 89 18

www.mdpi.com

Journal of Marine Science and Engineering Editorial Office

E-mail: jmse@mdpi.com

www.mdpi.com/journal/jmse